Dying for Work

Interdisciplinary Studies in History

Harvey J. Graff, General Editor

Dying for Work:

WORKERS' SAFETY AND HEALTH IN TWENTIETH-CENTURY AMERICA

edited by

David Rosner and Gerald Markowitz

INDIANA UNIVERSITY PRESS

Bloomington and Indianapolis

Chapter 4 originally appeared in *Science and Society,* Winter 1984/85. Chapter 6 originally appeared in *Journal of Social History* (Spring 1985). Chapter 8 originally appeared in *American Journal of Public Health* 75 (April 1985). We thank these journals for permission to reprint these articles.

Library of Congress Cataloging-in-Publication Data

Dying for work.

(Interdisciplinary studies in history)
Bibliography: p.
Includes index.
1. Industrial hygiene—United States—History—
20th century.   I. Rosner, David, 1947–
II. Markowitz, Gerald E.   III. Series.
HD7654.D95   1987   363.1'1'0973   86-10260
ISBN 0-253-31825-4

1 2 3 4 5 90 89 88 87

CONTENTS

For our children
BILLY, ELENA, AND TOBIAS MARKOWITZ
AND ZACHARY AND MOLLY ROSNER

Acknowledgments

A number of people were instrumental in the creation of this book. First, we would like to thank the editor of the series, Harvey Graff, for his enthusiastic encouragement of this effort to chart a new course in labor and health history. Harry Marks has given us the benefit of his insight and research. Lorin Kerr has, as always, provided enthusiastic support for this effort as well. Anthony Bale and David Kotelchuck, contributors to this volume, have also provided important contemporary perspectives to the historical analyses in a number of our own essays. The historians at Baruch College and John Jay College have also been helpful. We would especially like to thank Myrna Chase, Blanche Cook, Tom Frazier, Paul Le Clerc, and Bill Preston. Of course, Herbert Gutman, our colleague and friend, was an inspiration. We treasure his memory.

We would like to thank the Milbank Memorial Fund, and especially Sidney S. Lee and David Willis, for providing financial and intellectual support for this project. The Professional Staff Congress–City University of New York Research Foundation also provided financial assistance. Rona Ostrow and William Yu of Baruch College's Business Resource Center were also immensely helpful in the preparation of this manuscript.

We would especially like to thank our authors, who put up with our gentle and not so gentle prodding during the past year. They were truly a wonderful and collegial group with whom to work. Our families were, of course, supportive and loving.

Two people who deserve special thanks are Barbara and Paul Rosenkrantz. Barbara provided her special gifts as a scholar, colleague, and friend, and Paul lent a special warmth and kindness for which he is deeply loved. Finally, we would like to say that working on this project together was surely a joy for both of the editors.

David Rosner and Gerald Markowitz

Introduction: Workers' Health and Safety—Some Historical Notes

In recent years, American labor and health historians have examined the effects of industrialization on the lives of workers and their families. Labor historians have looked at the transformation and control of the workplace while health historians have examined the relationship of immigration, industrialization, and urbanization to the public's health. One arena in which these two fields share common ground is the study of workers' safety and health, a field that looks directly at the effects on Americans of changes in the workplace and environment. This book seeks to bring the discussion of workers' health and safety into the mainstream of American labor and health history.[1] We believe that the health status of workers and their families is central to the issues of workers' control and public health in industrializing America. We begin with the premise that the exploitation of labor is measured not only in long hours of work and lost dollars but also in shortened lives, high disease rates, and painful injuries.

The chapters in this book do not provide a comprehensive history of the field of safety and health. Rather, they are meant to illuminate interrelationships between industrial and social organization and workers' health. Along with labor and health historians, we have asked an economist, two physicians with expertise in occupational medicine, a sociologist, and a chair of an academic industrial hygiene program to address historical issues in safety and health in an effort to help define the parameters of this new endeavor. Here they cover four major areas of interest: The first section addresses the alternative models that workers, activists, industry, and government have developed for addressing issues of prevention of, and compensation for, the ravages of industrial accidents and diseases. The second section looks at the development of state and federal regulation of safety and health in the plant. The third section focuses on one of the most ubiquitous industrial poisons, lead, and addresses the political and scientific issues surrounding its control. The final section of this volume examines the social and economic conditions surrounding three devastating industrial diseases of the twentieth century: asbestos-related disease, byssinosis, and radium poisoning. The chapters, all drawn from the ongoing work of the authors, are case studies in the history of occupational safety and health, and it is hoped that this work will stimulate a more synthetic integration by other historians of labor and health.

What becomes clear from the work of the various authors is that occupational safety and health history illuminates the tensions among and within the scientific, economic, and political spheres in American society. In its most basic form, the struggle between labor and capital to control the means of production has set the context within which safety and health programs and policies have developed. The great struggles of labor in the late nineteenth and early twentieth centuries frequently revolved around wages, hours, and working conditions. As Alan Derickson, Robert Asher, Anthony Bale, and David Rosner and Gerald Markowitz show in their chapters on labor, big business, and government, the problem of safety and health was near the surface in many labor struggles. Labor organizers, for example, knew that underpaid, overworked, and poorly fed workers were more likely to be injured or incapacitated on the job: a miner who worked 12 to 14 hours a day could not stay alert enough to avoid injury from unguarded machinery in unlighted, noisy, and humid shafts; miners found their lives constantly threatened by speed-ups, explosions, dust, and suffocation. In contrast, employers, fearing the loss of control over production and the added costs of improving working conditions, have traditionally resisted reforms. Paternalistic benefit programs were often adopted by management as a means of increasing management's control over the work force.

The role of health professionals, most notably of doctors and industrial hygienists, has been shaped by this conflict between labor and capital. More often than not, professionals, even when they have sought to maintain their objectivity, have found themselves compromised by the highly political implications of their work. Many professionals have been able to find employment only with private industry and have often adopted the values and assumptions of their employers regarding responsibility for risk. Historically, only a small number have been hired by government or academe, thereby retaining some semblance of independence. But even there, as the chapter by Craig Zwerling shows, the professional cultures and political context of industrial hygiene and occupational medicine have often made them less than sensitive to the problems of the common laborer.

Most often, the struggles over safety and health have been fought out within particular plants, around specific occupational hazards or diseases. But increasingly, workers and public health professionals have turned to government, at both the state and the national level, for protection. This has moved the conflict to another level and has forced government officials to mediate between labor and capital. Although the issues often have been framed in technical, scientific language, their social and political impact has been clear. As a result, government's resolution of these disputes generally has had less to do with the "objective" or scientific evidence than with the relative power of the contending forces. Zwerling's chapter and those by Jacqueline Corn and Rosner and Markowitz illustrate the contradictory impulses embedded within the governmental response.

Most often, the battles over industrial conditions have been fought over particular workplace or environmental issues. Lead, a central industrial mineral, has continually been a focus of debate. Arguments over its relative importance or danger to society have been shaped by changing power relationships at different historical moments. The chapters by Rosner and Markowitz, William Graebner, and Ruth Heifetz all address the changing context of the debate. Two of the chapters specifically address the politics of leaded gasoline, an environmental and industrial toxin; the chapter by Heifetz looks at the role of women in identifying and defining the reproductive risks of lead.

There are other characteristic hazards associated with modern industrial society, and the three chapters by Angela Nugent, David Kotelchuck, and Charles Levenstein, Dianne Plantamura, and William Mass address three of them. Nugent looks at the politics of defining radium as a toxin in the 1920s. Kotelchuck, a practicing industrial hygienist, traces the early history of the asbestos tragedy. And Levenstein, Plantamura, and Mass look at the role of labor in the slow identification of brown lung disease following World War II. Together, the chapters give us a sense of the complex political history of industrial epidemiology.

The history of safety and health issues is particularly rich because it involves not only the contending forces of professionals, government, management, and labor, but also a panoply of journalists, social workers, consumer advocates, environmentalists, and industrialists. This concern over working conditions reached a peak in the first decade of the 20th century in the wake of the revolutionary social and economic changes that America had just undergone. In the decades after the Civil War, Americans witnessed the virtual explosion of urban and manufacturing centers. This was shocking to Americans reared in rural settings. In the antebellum era, most Americans lived on farms or in small towns; the factories that existed were scattered in mill towns and cities in the Northeast. With the growth of the transcontinental railroads, the development of national markets, the increased exploitation of natural resources such as coal and iron, and the massive immigration of labor from rural Europe to the growing industrial cities of the East and Midwest, conditions of work changed dramatically. America moved from being a fourth-rate industrial power to being the leading industrial producer in the world; and as some of the chapters in this volume indicate, working conditions for many laborers deteriorated. Speed-ups, monotonous tasks, and exposure to chemical toxins, metallic and organic dusts, and unprotected machinery made the American workplace among the most dangerous in the world. In mining, for instance, England, Germany, and France experienced fewer than 1.5 deaths per thousand workers during the first years of this century. In the United States more than three miners in every thousand could expect to die while working in a mine during any given year.[2] Clearly, the enormous wealth produced by the new industrial plants was achieved at an inordinate social cost. "To unprecedented prosperity . . .

there is a seamy side of which little is said," reported one observer in 1907. "Thousands of wage earners, men, women and children, [are] caught in the machinery of our record breaking production and turned out cripples. Other thousands [are] killed outright. . . . How many there [are] none can say exactly, for we [are] too busy making our record breaking production to count the dead."[3] In a theme that would be repeated, many compared the toll of industrial accidents to an undeclared war. As early as 1904 *The Outlook*, a mass-circulation magazine, commented on the horrendous social effects of industrialization. "The frightful increase in the number of casualties of all kinds in this country during the last two or three years is becoming a matter of the first importance. A greater number of people are killed every year by so-called accidents than are killed in many wars of considerable magnitude," it pointed out. "It is becoming as perilous to live in the United States as to participate in actual warfare." The editorial demanded that the state document the extent of industrial accidents "in order that the people of the United States may face the situation and understand how cheap human life has become under American conditions."[4]

Prior to the 20th century workers were barely protected by a variety of state and federal statutes that addressed a number of very specific workplace conditions.[5] However, during the progressive period, there were demands for a more systematic, more integrated approach to understanding and affecting the excesses of industrial capitalism. To reformers at the turn of this century, the country appeared woefully behind the industrialized European community.[6] And during the first decades of this century a movement arose that brought together a broad coalition of radicals, reformers, labor leaders, and even business representatives. Within the context of a larger social movement to reform health conditions in general, contemporary discussion of occupational safety and health came to be part of broader social concerns regarding workers' housing, sanitation, and general living conditions. Reflecting a broad ecological notion of the relationship between the environment and the health status of workers, some argued that there was no clear means of distinguishing occupational safety and health from other social and environmental problems.[7]

Representatives of the labor movement shared this broad conception of safety and health. At the 25th Annual Convention of the American Federation of Labor in 1906, for instance, the delegates identified tuberculosis as one of the organization's most pressing problems. In a dramatic chart showing the death rate from consumption in fifty-three occupations, the A.F. of L. pointed out that marble and stone cutters, cigar makers, plasterers, printers, and servants all had death rates well above four per thousand while bankers, brokers, and officials had the lowest death rates—below one per thousand. In an address before the convention one speaker spelled out the connection between work, wages, living conditions, and tuberculosis: "All this means, really, the regulation of factory conditions, the regulation of housing, and the passage of child labor laws" were essential for battling the "Great White Plague." Significantly, the A.F. of L. claimed that health status would be a direct measure of the success of the trade union movement. "In the same degree that the trade union movement becomes powerful will it establish such improved conditions that will check and eliminate the ravages of consumption."[8] Other unions noted that "no movement" that ignores the industrial workplace can mount an effective "campaign against tuberculosis."[9]

Union campaigns to make shop conditions more sanitary were linked to broader public health issues—most importantly, to the battle against infectious disease. Cleaning up the workplace and keeping the work force healthy were seen as benefits to both the worker and the public. In 1910 the greater New York local of the International Union of Bakers and Confectionery Workers conducted a successful strike to demand more sanitary working conditions. "Perhaps no phase of the trade union movement has ever affected the public so directly as the agitation for sanitary conditions in the bake shops," commented one leading periodical. In May of 1909, 3,000 Jewish bakers struck, and less than a year later 4,000 German workers followed suit.[10] The union identified unsanitary workshops and the spread of infectious disease with non-union bakeries and sought to make an alliance with the public by linking unsanitary working conditions, the health of the public, and unionization: "the great bread eating public of this country should see to it that the bread they eat bears the bakers' union label."[11]

The very prevalence of contagious diseases such as diphtheria, influenza, tuberculosis, and typhoid within the working class spurred more middle-class consumers' groups to take up the issue of health conditions on the job and in the home. In part the appeal was based on fear, to make sure the middle class and wealthy would not be infected by goods tainted by sick workers. In the growing garment industry of New York, many dresses, shirts, and trousers were sewn on a piecework basis in tenement slums, raising the specter that the same diseases infecting those in tenements would be transmitted to the men, women, and children of the middle class. It was this terror that led the National Consumers League to become active in tenement reform and anti-tuberculosis campaigns. Its label, along with the International Ladies' Garment Workers' Union's, came to be the mark of clothing manufactured under hygienic conditions. The League became involved in a wide variety of labor related issues including fire safety, workers' compensation, and occupational disease legislation.[12]

The special social conditions surrounding work at the turn of the century led to a broad conception of the meaning of occupational safety and health—and to seemingly incongruous alliances. The rapid unregulated growth of industry, the enormous immigration of foreign workers, the growing strength of the Socialist Party, the fear of social unrest, and the terror engendered by infectious disease gave the movement a special appeal.

Occupational safety and health was part and parcel of a larger movement to reform American society; it contained all the contradictions alliances of this larger social movement. At one end of the spectrum were radicals such as the famous feminist socialist Crystal Eastman and the socialist head of the Amalgamated Clothing Workers Union, Morris Hillquit. At the other end of the political spectrum were the "enlightened" owners of corporations such as International Harvester and United States Steel who sought to divert the movement toward voluntary welfare and safety programs. Between these two extremes were a wide variety of middle-class reformers and conservative labor representatives who recognized that uncontrolled capitalism was killing and maiming so many workers that it was undermining the legitimacy of capitalism itself.

On a concrete level, the movement was successful in achieving a few important legislative and political victories. Perhaps the most important in terms of the long-term battle to control workplace hazards was the effort to get the federal and state governments involved in regulating working conditions and the passage of workers' compensation laws. In the decade after 1911, 25 states passed laws that guaranteed some form of financial compensation to workers injured on the job and to their families. Employees gave up the right to sue an employer in return for prompt and sure remuneration for an accident incurred on the job, regardless of who was at fault. Whatever the criticisms that we may today have of this system, at the time it was considered a victory for those seeking to improve working conditions. The passage of workers' compensation, in combination with other factors (such as the decline of the Progressive movement in general and the onset of World War I), also resulted in a dramatic change in both the ideology and the program of safety and health reformers. Central to the Progressive movement was a need to integrate safety and health issues into broader social and political struggles around the responsibilities of the larger society to industrial workers and their families. With the decline of this larger social movement, safety and health was relegated increasingly to businessmen themselves. Under the slogan "Safety First," the scope and focus of the movement narrowed dramatically.[13]

Workers' compensation gave businessmen an interest in reducing injuries, because each company's insurance premiums would be based on its own accident rates. Corporate leaders, still conscious of the significance of popular opinion in the creation of workers' compensation laws, also feared that future reform efforts might be more radical. Hence, they sought means by which to gain greater control over the movement to make the workplace safer. The National Safety Council, organized in 1911, became the focus of most business efforts around workplace safety. Its general proposition was that "safety pays" and that employers benefited economically by taking the lead in accident prevention.

The new safety movement of the late teens and early twenties can be distinguished from the older Progressive Era efforts in a number of ways. First, the new business-led efforts emphasized the responsibility of the workers themselves, rather than that of industry, to prevent accidents. Second, they narrowly defined the problem as one of safety rather than health. Third, they saw professionals, rather than workers or reformers, as the prime source of change and thus sought to take the discussion of safety and health out of the public arena.

The effort to blame the workers built upon the general antipathy for immigrants that prevailed at the turn of the century. One publicist maintained that the importation of cheap foreign labor was responsible for the high accident and disease rates that prevailed in American industry. Citing the high percentage of foreign-born workers who suffered from lead poisoning in New York City, Arno Dosch explained it as a product of the immigrants' ignorance: "The Americans know how to take care of themselves. . . . They wash their hands and faces when they stop work." He held that "the immigrants from Eastern Europe do not unless someone stands over them and makes them do it." He went beyond this observation to assert that "this class of Eastern European peasant lacks the intelligence and initiative either to avoid the ordinary dangers of rough labor or to keep in efficient health."[14]

Most "safety first" advocates saw the problem of worker carelessness more broadly. Rather than condemn only the immigrant for his recklessness and ignorance, they blamed his American counterpart as well. Whereas the progressives held that industry was responsible for most accidents because of its failure to provide safeguards from dangerous machinery, the "safety first" movement turned the problem on its head. They claimed that few accidents were due to faulty machinery or inadequate safeguards and that most were the fault of workers themselves.[15] One movement member maintained that "one of the principal sources of accidents is the worker himself. Carelessness, thoughtlessness and lack of knowledge all conspire to cause him injury."[16] The *National Safety News*, official organ of the National Safety Council, was more specific in its analysis of the causes of accidents. While "faulty plant conditions" was thirteenth on its list, the major causes were "ignorance of English," "inexperience," "mental limitation," "faulty attitudes—fatalism, antagonism toward the industrial medical department, timidity," "mental sets—the chronic kicker, the grouch and the radical have mental sets," and "excitability." Last on their list was "fatigue," but the author was quick to point out that this was "not necessarily from overwork, but from hard and stressful living conditions, undernourishment, loss of sleep, financial and domestic worries or emotional stress."[17]

Industry argued that improvements in plant safety required two major programmatic changes: First, and most importantly, workers had to be educated to protect themselves from accidents. Second, this responsibility had to reside with experts who understood the newly emerging field of industrial medicine within the context of the overall needs of industrial production. For "safety first" advocates the workplace was no longer seen as part of a larger social and economic environment. Nor was the workplace seen as being in need of radical or substantial reorganization as defined by an earlier generation of progressive reformers. The first objective, that of affecting the work force, was to be handled by the newly emerging field of occupational medicine, and specifically by the company physician working in the plant. "The point of approach to the human potential had best, therefore, be through the industrial dispensary. Under a high-grade physician it will be the great melting pot of the human experiences of men. Here the virtues and the weaknesses of the men will be most apparent. The physician will also be confessor, advisor, priest. Through him the employee may learn that it pays to be healthy, steady, and of good habits. He does not hesitate to preach the 'Sober First' campaign."[18] During the 1920s, hundreds of companies hired their own physicians, nurses, and other engineering and medical personnel. The control of workplace hazards came, then, to be dominated by experts in the newly emerging fields of industrial hygiene, industrial medicine, and engineering and lost its close identification with the broader perspectives of progressives and labor reformers.

Until this point, with the exception of World War I, nearly all governmental regulation of occupational health and safety was carried out at the state level. The federal role in controlling health hazards was extremely limited. This meant, in practice, that there was little or no uniformity among safety codes, statutes, or enforcement practices. Even within states, codes often varied greatly. The strength of state and local codes and enforcement practices varied between urban industrial and rural agricultural settings, between north and south, and, most importantly, between strongly organized and non-union states.

The problems of this highly fragmented and weak governmental presence were brought into high relief by a major disaster that affected the health of at least two thousand predominantly black workers in the anti-labor state of West Virginia in the early years of the Depression. In the spring of 1933, at the start of the New Deal, a series of lawsuits were filed revealing that over the course of the past three years 476 workers had died and most of the other workers were disabled by acute silicosis caused by the inhalation of dust while drilling 3.75 miles of tunnel to divert water from New River to a hydroelectric plant at Gauley Junction, West Virginia. The hydroelectric plant was being constructed by New Kanawha Power Company, a subsidiary of the Union Carbide Company, to provide power to its nearby petrochemical plant. Despite the fact that the company knew that they were exposed to dust that was 97 to 99 percent pure silica, no precaution was taken by management to provide the workers with masks, ventilation or any other protection. The legacy of the lack of government regulation and the dependence of workers on management paternalism was revealed in congressional testimony that reported that company physicians systematically misled workers who complained of lung ailments: "the company doctors were not allowed to tell the men what their trouble was. "A Dr. Mitchell of Mount Hope, a company doctor, testified" . . . that he had told the men they had 'tunnelitis'." Despite denials by the company of knowledge of the dangers, it became obvious that management and professional alike took precautions to protect themselves from the dangers of silicosis by routinely using face masks denied workers when entering the tunnel. Furthermore, it was revealed that company geologists had determined before construction of the tunnel that the mountain through which they were boring was almost pure silica: They shipped the silica to another Union Carbide subsidiary "where it was stored in the yard. It was so pure that it was used without refining." The horror of this event was brought to public attention when it was discovered that 169 of these workers had been buried in a mass grave in a nearby field by a local undertaker who was paid by the company itself.[19] This, the worst industrial disaster in American history, brought to public attention the need for increased government control over working conditions. During the New Deal administration of Franklin D. Roosevelt, Frances Perkins's Department of Labor engaged in a fascinating experiment in governmental involvement in safety and health (see chapter 6).

The postwar era witnessed a growing interest in the effects of industry in creating environmental pollution. Despite the waning activity of government in the Cold War years of the 1950s, a newer perspective on the relationship between industry and society developed in conjunction with a growing popular concern over environmental pollution and occupational hazards. These dual movements had very different origins and political perspectives. Many in the environmental movement had little or no interest in workers' problems and often found themselves at loggerheads with workers whose interests were substantially different. Often, environmentalists such as Rachel Carson, whose book *Silent Spring* popularized concern over DDT and other industrial products, took a decidedly anti-industrial stance that equated pollution with industrialization itself. This sometimes put them in conflict with labor, whose goal was to find ways of making industry safe, rather than to eliminate it altogether. Other environmentalists such as Barry Commoner worked more closely with labor, seeking ways of promoting environmentally sound policies that protected people both inside and outside the factory.

There was a similar divergence among professional industrial hygienists and labor leaders about the proper approach to occupational safety and health. In the fifties, professionals focused on the problems of plant safety rather than on longer-term health hazards. The few professionals such as William Hueper who had a broader perspective on the relationship between industrial chemicals and cancer found their professional lives cut short and their reputations sullied. But in the 1960s, activists such as Anthony Mazzochi of the Oil, Chemical and Atomic Workers' Union and Lorin Kerr of the United Mine Workers of America pressed their campaigns for comprehensive protection both for workers on the job and for the broader community. This new movement gained strength from the legitimacy of social activism provided by other contemporary movements for civil rights and the War on Poverty and against the Vietnam War. While these movements had limited appeal to many workers, some unions—the United Automobile Workers, hospital workers, the United Mine Workers and others—sought to forge ongoing alliances around common issues. Many other workers supported the demand for health and safety legislation because of the dramatic increase in injuries associated with the speed-up in production necessitated by America's growing involvement in Vietnam.

The most dramatic results of these movements were three pieces of legislation passed in 1969 and 1970. In 1969, Congress enacted the Mine Safety and Health Act, which provided protection to miners on the job and financial and medical assistance to those diagnosed with job-related diseases such as Black Lung. In 1970, Congress increased government protection of American workers through its passage of the Occupational Safety and Health Act, which mandated the creation of both the Occupational Safety and Health Administration (OSHA) in the U.S. Department of Labor and the National Institute of Occupational Safety and Health (NIOSH) in the Department of Health, Education and Welfare. OSHA was to set and enforce national standards for safety and health on the job while NIOSH, the research arm, was to establish safe levels of exposure to industrial pollutants. In the same year Congress passed the Environmental Protection Act, which created the Environmental Protection Agency (EPA) to enforce a wide range of programs aimed at reducing the dangers of pollution to the broader community. It may appear to many that this rush of legislation was a result of the movements that arose during the activist 1960s. We hope that this necessarily cursory introductory essay, along with the other chapters in this volume, helps in bringing the rich history of safety and health to the attention of historians, professionals, and workers alike.

NOTES

1. Other historians have investigated different aspects of occupational safety and health. William Graebner and George Rosen, for example, have written detailed work on various aspects of mine safety; Barbara Sicherman has recently published a collection of Alice Hamilton's letters; Paul Brodeur has recently published a volume on the history of asbestos-related diseases, and others have done detailed studies of specific industries or particular diseases. While all this work is extremely valuable,

only Henry Selleck and Alfred Whittaker in their encyclopedic work during the early 1960s, *Occupational Health in America,* (Detroit: Wayne State University Press, 1962), have sought any type of synthesis. See also Ludwig Teleky's *History of Factory and Mine Hygiene* (New York: Columbia University Press, 1948) for an overview of this subject; also, Judson MacLaury, "The Job Safety Law of 1970: Its Passage Was Perilous," *Monthly Labor Review,* March 1981:18–24; and Stuart Kaufman and Judson MacLaury, "Historical Perspectives" in *Protecting People at Work, A Reader in Occupational Safety and Health,* ed. Judson MacLaury (Washington: U.S. Department of Labor, 1982).

2. B. Reeve, "The Death Roll of Industry," *Charities and the Commons* 17 (1907): 791.

3. Ibid.

4. Editorial, "Slaughter by Accident," *The Outlook* 78 (Oct. 8, 1904): 359.

5. George M. Kober and Emery R. Hayhurst, eds., *Industrial Health* (Philadelphia: P. Blakiston's Son and Co., 1924), pp. vii–lviii.

6. See for example Alice Hamilton, *Exploring the Dangerous Trades* (Boston: Little Brown and Company, 1943), for a brief discussion of this interesting moment in public health history.

7. "The Tuberculosis Fight in an Industrial City," *Charities* 13 (Dec. 17, 1904): 279; Frederick Almy, "Transcript of Proceedings," *Second Report of the New York Factory Investigating Commission,* 1913, vol. 4, pp. 1831–32; and Graham Taylor, "Industrial Basis for Social Interpretation," *The Survey* 22 (Apr. 3, 1909): 9.

8. "How to Prevent Consumption," *The International Woodworker* 16 (May 1906):137–9; "Paul Kennaday on Tuberculosis," *The International Woodworker* 16 (May 1906): 139–40.

9. "Labor is Against Tuberculosis," *The Glassworker* 6 (April 1909): 6.

10. "A Strike for Clean Bread," *The Survey* 24 (June 18, 1910):483–88.

11. "Investigations Have Disclosed the Fact that Unhealthy and Poisonous Bread is Made in Non-Union Bake Shops," *The Woman's Label League Journal* (June 1913):13.

12. Charles Swan, "Enterprise Liability for Industrial Injuries," *Annals of the American Academy of Political and Social Science* 38 (July 1911):262–3; "The Consumers League Label and Its Offspring," *The Survey* 32 (Aug. 8, 1914): 478; Mary H. Loines, Chairman of the Brooklyn Auxiliary of the Consumers League of the City of New York to Hon. Robert F. Wagner, Dec. 12, 1912, *Second Report of the New York Factory Investigating Commission,* 1913, vol. 2, 1330–31, and vol. 4, 1576–77.

13. Donald Wilhelm, "Safety First: The New Social Work," *The Outlook* 107(July 25, 1914): 701; C. W. Price, "Some Outstanding Facts in the Safety Movement," *American Labor Legislation Review* 10(1920):26.

14. Arno Dosch, "Our Expensive Cheap Labor," *World's Work* 26(Oct. 1913):699.

15. "Some Causes of Accidents," *Scientific American* 124(April, 1921):352; "A study of the causes of 220,707 accidents which have occurred in the plants of the United States Steel Corporation shows that but 4.94 percent of the total number, excepting those in connection with overhead electric cranes, were due to machinery causes."

16. Fred G. Lange, "Safety and Accident Prevention," *Industrial Management* 61(April 1, 1921): 257.

17. W. R. Woodbury, *National Safety News* 7(May, 1923):25; see also Frank Moffett, Address to the National Safety Congress in "The Personal Element in Safety," *Literary Digest* 68(Jan. 8, 1921):102. There were limitless possibilities for blaming the victim: "Aside from the man's psychological make-up we may consider some of the secondary factors. The man may be too daring, and may like to run risks; he may be old or he may be young; or he still may need discipline; any one of these

coupled with the fact that he may have something on his mind or troubles of his own."

18. Otto P. Geier, "The Human Potential in Industry," *Scientific American Supplement* 84(Dec. 22, 1917):386.

19. U.S. Congress, House Committee on Labor, "An Investigation Relating to Health Conditions of Workers Employed in The Construction and Maintenance of Public Utilities," Jan.–Feb. 1936, in *West Virginia Heritage* (Richwood, WV: West Virginia Heritage Foundation, 1972), vol. 7.

Part One

Labor, Management, and the Legal System: Alternative Approaches to Protecting and Compensating Injured Workers

Workers' compensation has been the central focus of a number of studies of occupational safety and health history. Historians have detailed the inadequacies of the protection offered by the compensation system. It has been critiqued for its limited coverage of occupational disease, its role in eliminating the courts as a source of redress by injured workers and its use by industry to obscure management's responsibility for industrial accidents and illnesses. In this section the authors look at the issue of protection from a different vantage point. Alan Derickson considers the role of unions themselves in assuring the health and safety of their members. He examines a worker-sponsored hospital in the mining town of Coeur d'Alene, Idaho, in the 1890s and shows how safety issues were central to workers' demands and programs. Robert Asher looks closely at business-sponsored attempts to provide protection and welfare benefits to workers in a variety of industries in the late nineteenth and early twentieth centuries. He concludes that such initiatives were not motivated by a desire to aid workers and their families during time of distress but rather were a means of undermining worker- and union-based institutions that developed around safety and health. Not surprisingly, these business benefit plans afforded the company greater control over the work force by making workers dependent on the goodwill of the company, rather than on their union. Anthony Bale traces the crises within capital and in the legal system at the turn of the century that gave rise to the workers' compensation system. He points out that as the country emerged from its rural, agrarian past, social perceptions as to responsibility for workplace accidents affected the outcome of liability cases to an unprecedented extent. This created a crisis as some workers received extremely large settlements from companies and the legal system found itself overburdened. Finally, David Rosner and Gerald Markowitz look at an attempt by pro-labor activists during the 1920s to confront the health problems that were not being addressed by the recently adopted workers' compensation system. Organized by three radical women, the Workers' Health Bureau sought to address safety and health as a class issue that demanded rank-and-file activity rather than as a technical or scientific problem to be dealt with by professionals alone.

The four essays in this section suggest the importance of safety and health as integral to labor and management's struggles over control of the workplace. Together, these essays point to the importance of sponsorship in defining the limits of health and welfare initiatives and the alternative uses to which they can be put.

CHAPTER **1** *Alan Derickson*

"To Be His Own Benefactor": The Founding of the Coeur d'Alene Miners' Union Hospital, 1891

In the late nineteenth century, breakneck industrial development exacted from American workers an immense toll in illness, injury, disability, and death. Even more poorly understood than the full extent and nature of this human carnage are the self-help efforts of workers to cope with and to curtail their victimization under industrializing capitalism. In metal mining, one of the most hazardous of all industries, the grassroots health programs of local miners' unions grew in this period to encompass not only the provision of sickness and funeral benefits, lay nursing, and other traditional forms of mutual assistance but also the unprecedented creation of general hospitals in mining communities. In the 1890s and 1900s, hardrock miners through their unions planned, built, and controlled more than twenty hospitals in the western United States and Canada.[1]

In 1891 the lead and silver miners of northern Idaho established the first of these institutions, the Coeur d'Alene Miners' Union Hospital. The Miners' Union Hospital served two basic purposes for the miners and their unions. First, it delivered badly needed health care and thus provided an alternative to dependence upon employer-controlled medical programs. Second, together with other union mutual-aid activities, this institution forged greater solidarity among the organized workers and generated considerable public support for the miners' unions as benevolent organizations. After discussing the problems which impelled the miners to undertake this project, this chapter will focus on one critical episode in the campaign to establish the Coeur d'Alene hospital, the strike at the Bunker Hill and Sullivan mine in August 1891. The Bunker Hill dispute reveals the ways in which the very existence of the Miners' Union Hospital challenged important employers' prerogatives. Thus, the founding of the hospital was an integral part of the miners' overall defensive strategy to resist the sweeping deleterious changes in the social relations of producing metals brought on by the rise of the corporation and the acceleration of industrialization.

The development of hardrock mining in the Coeur d'Alene area reflected the national pattern of explosive growth in this industry in the decades after 1860. For the United States as a whole, silver output advanced from 1,547 ounces in 1861 to 58,330 ounces in 1891, and lead output increased from 14,100 tons to 198,363 tons over the same period. The discovery of rich deposits of lead and silver in the mid-1880s rapidly transformed a remote wilderness in the "panhandle" of Idaho into one of the world's leading mining centers. In 1889, Shoshone County, Idaho mines yielded 18,564 tons of lead ore, making it the nation's third leading lead-producing county. The value of all ore extracted in the district in 1890 was estimated to be approximately $10 million.[2]

The corporation was the only form of business organization capable of fully exploiting these resources. Big capital was necessary to acquire and consolidate mining claims, engage in protracted litigation over the claims, retain a sizable and relatively well-paid labor force, buy and maintain mining and milling equipment, and pay stiff railroad shipping rates.[3] To take one key example, upon purchasing the Bunker Hill and Sullivan mines for $650,000 in 1887, Simeon G. Reed incorporated the Bunker Hill and Sullivan Mining and Concentrating Company for $3,000,000 and commenced construction of a 150-ton-per-day ore concentrator and other major capital improvements. With eastern, western, and midwestern capitalists investing heavily in the district, absentee ownership became predominant.[4]

With the mining boom a scattering of small towns grew up around the mines along the Coeur d'Alene River and its tributary creeks. In June 1888 the *Wardner News* announced that "experienced miners from all directions are commencing to appear," hailing the arrival of "that class most desirable in a mining country."[5] Veteran mining craftsmen converged on the Coeur d'Alene from the Comstock Lode of Nevada; Butte, Montana; and other mining areas. By 1891 the district employed more than 1,200 mine workers.[6] The fragmentary evidence extant indicates that a substantial proportion, perhaps a majority, of the hardrock miners were immigrants and that an overwhelming majority were unmarried men. The largest immigrant groups were the Irish and Canadians, followed by the Germans, Swedes, and Italians. Also present was the indispensable contingent of Cornishmen who, drawing on centuries of metal-mining experience in the old country, provided most of the traditional knowledge of extracting ore on the western frontier.[7]

The pre-industrial craft of mining metals required considerable mental and physical skills. Virtual masters of their workplace, miners customarily determined how to pursue meandering veins of ore through a mountain, which tools and materials to use, what pace of work to maintain, and how to safeguard working conditions. Subsumed under this wide-ranging autonomy was a myriad of daily judgments which were never routine because each work situation was unique. Common decisions, such as assessing the stability of the rock "roof" overhead or troubleshooting a misbehaving fuse, could mean life or death for the miner, his co-workers, and his employer's investment. In the era before inanimate sources of power to perform work became predominant, the miner's basic manual tasks were hand drilling, loading the holes drilled with blasting powder, and igniting the charges. By far the largest share of work time was spent driving a steel drill through solid rock with a four- or eight-pound hammer. Like so many other duties, hand drilling required strength and dexterity. Former miner Frank Crampton remembered performing this arduous task "from every position, excepting standing on one's head—in all directions—up, down, at an angle, or at one side." In a sense, at the face of the mine the miner literally created his own workplace by continually blasting out new space, erecting timber roof supports, digging drainage ditches, and doing other maintenance chores. Thus, skilled workers controlled the process of producing metal-bearing ores.[8]

In the closing decades of the nineteenth century, hardrock mining became increasingly industrialized. With the successful application of first steam and then electric power to diverse underground operations, mining technology was transformed in ways that both altered traditional work practices and posed unprecedented challenges for workers' control of the workplace. In particular, the advent of machine drilling after 1870, coupled with the adoption of dynamite as a blasting agent, brought about a crisis for craft methods of production. Power drills penetrated rock many times faster than did hand drills. Dynamite dislodged far more rock per explosion than did black powder. Contemporaneous advances in ore milling technology further facilitated the emerging strategy of relying primarily upon expanded output and economies of scale to achieve efficient production, rather than on the more precise but less productive craft methods involving a more "surgical" removal of ore.[9]

The new technology not only placed less value on very high levels of skill but also required fewer skilled workers. Because ore loading and hauling experienced no corresponding technological breakthroughs in the Gilded Age, the occupational structure underground shifted substantially. Prior to the onset of industrialization, a division of labor had already been created between the skilled miners* who drilled and blasted, the muckers who shoveled the broken rock into ore cars, and the carmen who pushed the loaded cars through the mine. The immediate impact of industrial methods was to increase markedly the proportion of unskilled muckers and carmen in the underground work force and to intensify greatly their work through simple speedup. Thus, when power drills were introduced into the Coeur d'Alene mines in the late 1880s and early 1890s, many skilled miners were driven down into unskilled jobs away from the mine face. The changing balance of power between employers and workers, together with declining metal prices due to chronic world overproduction, inevitably led to mounting efforts to cut wages.[10]

Metal miners responded to the growing power of capital by forming

---

*Under a strict definition of the term, only those craftsmen who worked at the mine face were considered to be miners. However, unless otherwise specified, this essay uses "miner" to denote any mine worker.

and/or strengthening industrial unions. Throughout this period many, if not most, local miners' unions were organized in reaction to a wage cut or the imminent threat of one, beginning with the Comstock Lode organizations in the 1860s. In the pre-industrial phase of mining when the preponderance of underground employees were craftsmen, the unions established an egalitarian wage policy, advocating a uniform wage, usually $3.50 or $4.00 per day, for all underground workers. As the development of the industry forced a growing share of workers into unskilled occupations, the miners' unions clung tenaciously to the equal-pay principle. In 1887 the first union in the Coeur d'Alene, the Wardner Miners' Union, was organized after the Bunker Hill and Sullivan ended uniform pay by reducing the wages of carmen and shovelers. Independent miners' unions, taking in both skilled and unskilled workers, were organized in the other three principle mining communities of the Coeur d'Alene—Gem, Mullan, and Burke—and the Wardner union was reorganized in the fall of 1890. By mid-1891 these unions had executed a series of strikes which won the $3.50 rate for all underground workers at all the district's mines except the Bunker Hill.[11]

Health problems, of both occupational and nonoccupational origins, were also critical factors not only in the founding but also in the ongoing functioning of the Coeur d'Alene miners' unions, just as they were for miners' organizations throughout the western United States and Canada in this period. Unions institutionalized formally the long-held miners' customs of mutual aid to sick and injured co-workers and respectable burial of the dead. These customs followed naturally from deeply ingrained work habits of mutual protection underground. Miners, especially immigrants without kin in this country, were painfully aware of the demeaning alternatives to self-help—charity, the poorhouse, and the potter's field. Hence, the high priority accorded mutual assistance reflected firmly held values of independence and fraternity which grew from basic needs. In this aspect of the development of the mining labor movement, the Comstock unions served as prototypes, administering disability and death benefit programs beginning in the 1860s. When in the preamble to their constitution the Coeur d'Alene miners declared their intention to "enable the miner to be his own benefactor," they directly borrowed the language of the Virginia City, Nevada Miners' Union constitution of 1879.[12] The Idaho unions were also influenced by the successful programs of the Butte Miners' Union, which not only paid the usual sickness and death benefits but also administered a special hospital benefit fund. Similarly, another important early organization, the Lead City Miners' Union, was incorporated in South Dakota in 1880 as a benevolent organization. In general, mutual assistance was nearly as important as the wage issue for the hardrock unions, and it helped to promote the solidarity necessary for the defense of wage standards.[13]

The Wardner Miners' Union and the other Coeur d'Alene unions administered disability and death benefits and augmented this financial assistance with strong social support to needy members and their families. The fraternalism which imbued this program was expressed in a union statement in 1891: "We care for our sick and alleviate the pains of our injured with a brotherly spirit. Our dead are laid to rest with all the respect and propriety becoming such sad occasions." Using members' dues and initiation fees, the unions paid sickness and accident benefits of ten dollars per week for up to ten weeks and granted up to five additional weeks' benefits in cases of "great bodily injury."[14]

The Coeur d'Alene unions made considerable efforts to honor their dead. The unions paid the funeral expenses for deceased members, noting with pride that they did not "find a pauper's grave as in the past." Union halls were frequently used as funeral parlors. In the case of Charles Wright, killed in the Poorman mine in April 1891, the body of the deceased lay in the Burke Miners' Union hall for viewing before it was buried "with miners' honors." As was common among hardrock unions, the Coeur d'Alene constitutional bylaws required the union president to conduct a funeral for any member who died without next of kin in the area. The rank-and-file membership attended the funerals en masse. The long processions to the Miners' Union Cemetery outside Wallace and the graveside ceremonies there clearly were important rituals which both reaffirmed and deepened miners' loyalty to one another.[15] These activities contrasted sharply with some employer-run programs of this period, such as the disposal procedures at the Treadwell gold mine on Douglas Island, Alaska, where it was reported that "if they killed a man (and the statistics show they have averaged a man a day for the last ten years [up to 1908]!) he was wrapped in canvas and carried away to the cemetery and buried by the company and by company men, not giving his relatives or the shift a chance to lay off in respect to [sic] their brother or fellow-workman."[16]

In January 1891 the four miners' unions in northern Idaho formed a district-wide federation which was officially named the Central Executive Miners' Union of Coeur d'Alene, but known most commonly as the Coeur d'Alene Miners' Union. The central purposes of this organization were to coordinate more closely the miners' efforts to attain and defend the $3.50 wage and to allow their unions to enter into joint undertakings. Immediately upon its formation, the Coeur d'Alene Miners' Union initiated as its first major project the establishment of a union hospital.[17] Beyond its instrumental value in uniting their labor movement and in promoting a public image of the new organization as a charitable body, the hospital was conceived primarily in response to the occupational hazards of mining and the inadequacy of existing health care arrangements.

Working conditions in the developing metal-mining industry were extraordinarily dangerous. In his study of the western miners of this period, Richard Lingenfelter estimated that one out of every eighty miners was killed on the job each year and that one in thirty suffered a disabling injury. Among the innumerable safety hazards were roof cave-ins, falling rocks, unexpected dynamite explosions, underground fires, faulty hoisting apparatus, unguarded shafts and ore chutes, and other unsafe equipment and procedures. Health hazards included exposure to silica, lead, zinc, cadmium, arsenic, radiation, carbon monoxide, extremes of heat and cold, and a host of ergonomic risk factors. Although no purpose is served by romanticizing preindustrial working conditions, it is clear that the miners themselves believed that industrialization exacerbated some pre-existing risks and generated new ones. For example, Coeur d'Alene miners recognized that power drills led to increased lead poisoning and to the adverse health effects of overwork for ore shovelers. No mention was made, however, by any union hospital advocate of the worst hazard of machine drilling—i.e., markedly elevated exposure to silica dust. This hazard produced an epidemic of silicosis, a chronic respiratory disease, then commonly known as "miners' consumption." In a sense, this oversight is not at all surprising, given the long latency period of this insidious disorder and the consequent improbability that the miners would establish a hospital to anticipate a problem whose full impact was years away. In any case, the myriad acute occupational health and safety problems, combined with substantial non-occupational problems of infectious disease, provided an ample rationale for building a hospital.[18]

Particularly compelling was the growing number of severe and fatal mine accidents. In March 1889 the *Coeur d'Alene Sun* recounted "A Mine Tragedy" at Mullan in which John Murphy was "badly cut and bruised" and John Waters was killed, his head "literally torn to pieces," by a premature blast. In February 1890 an avalanche smashed the boarding house at the Custer mine, killing six.[19] When in September 1890 a box of blasting caps exploded in George Peters' hands, it was found that "a large portion of the left hand, together with pieces of fuse and caps, had been blown into the abdomen." In October 1890 two more Coeur d'Alene miners lost their lives through accidents on the job.[20] Thus, a mounting toll of death and disability created momentum for improved health care services and facilities.

The miners sharply criticized obvious deficiencies in employer-administered health care arrangements. The limitations of the prevailing system and the miners' attitude toward it were succinctly summarized by John Sweeney, president of the Wardner Miners' Union: "[O]ne dollar per month was deducted from each man's pay as a hospital fee, but no hospital was provided. This money was paid over to a doctor, who gave medical attendance when called upon. This plan was very unsatisfactory to the miners. . . ."[21] Apparently all the major mine operators in the district participated in this plan, under which a physician contracted to provide services. By 1891 this mandatory payroll deduction generated over $1,000 per month from the miners, but the services were considered to be almost worthless. By establishing their own hospital and hiring their own physicians, the miners simply sought to improve upon an inadequate program which did not deliver what they deemed to be full value for their money.[22]

Because it was employer-controlled and perceived to be part of a larger strategy of social control, the company doctor system was objectionable to the miners as a matter of principle. They resented their exclusion from the process of selecting their physician and believed that the company doctor felt responsible only to the mine owners who had hired him, not to the workers who paid his salary. Although no specific allegations of malpractice were made against any of the company physicians of this period, the actions of one of the men appointed to this position supported the miners' criticisms regarding subservience to the employers. When martial law was declared in the Coeur d'Alene during a violent strike in 1892, Dr. William Sims demonstrated a zealous willingness to serve the mine operators. Appointed acting sheriff of Shoshone County, Sims personally identified numerous union activists for arrest by the National Guard. At that time the *Spokane Review* described the employers' physician as a man "cordially hated by every union man in the county."[23]

In addition, the Coeur d'Alene miners opposed the company doctor as one representative of a corporate system which threatened their personal freedom in new areas outside the employment relationship. In particular, they viewed with alarm the spread of company stores, bunkhouses, and boarding houses, which they were explicitly required or implicitly coerced to patronize.[24] According to James Sovereign, editor of the *Idaho State Tribune* in Wallace (and former general master workman of the Knights of Labor), "The employee who does not trade at the company store usually . . . is soon discharged from the service of the company for some imaginary cause."[25] The miners protested against all these institutions as inimical to their right to spend their earnings as they pleased. Thus, the assertion of a right to independence from comprehensive corporate domination was a factor in the union hospital campaign.[26]

Following the decision to build a hospital, plans for constructing and financing the institution proceeded under the direction of a committee composed of two representatives of each of the four unions involved. The committee reviewed and submitted to the constituent unions two alternative proposals for sites. By referendum the union members selected the centrally located town of Wallace as the location for the hospital.[27] While a citizens' committee conducted a fund-raising campaign among the general public, the unions committed themselves to allocate money from their respective treasuries, levy a special assessment on each union member, and increase their membership (and hence their financial base). In addition, the central union raised over $400 from a Saint Patrick's Day benefit dance for the hospital fund.[28]

In April 1891 a meeting between union representatives and mine owners resulted in an agreement on an appropriate role for employers in the project. A bilateral governing board for the hospital was constituted, consisting of five members—one from each of the four unions and one from the mine operators' association. In return for participation in governance, the employers agreed to remit automatically to the union hospital the dollar per worker per month being deducted for the company physician. Thus, the unions gave up only a modicum of policy-making freedom in exchange for a large measure of financial security. Soon every operator in the district except Bunker Hill and Sullivan had agreed to support the new institution. Bunker Hill persisted in making a compulsory deduction and retaining a local physician.[29]

While plans for a larger permanent building were being drawn up, the Coeur d'Alene Miners' Union Hospital, probably the first union hospital in the United States, began operation on May 8, 1891, in temporary quarters. Facilities included a ward of six beds, an operating room, and a dispensary "with enough chemicals in sight to furnish bane or antidote for any number of people in any man's country." The initial staff consisted of a superintendent, two physicians (part-time), and a nurse. In June 1891 the unions decided to transfer the day-to-day administration and nursing responsibility for the hospital to the Sisters of Charity of Providence, a Roman Catholic order which operated a number of hospitals in the Pacific Northwest. The unions remained financially responsible for maintaining the temporary facility and erecting a more substantial permanent one.[30]

On June 29 a union committee presented Bunker Hill general manager Victor M. Clement with a petition signed by a majority of the firm's mine workers, which requested that the company take part in the new hospital plan: "Your employees respectfully ask that one dollar per month be deducted from each man's pay and turned over to the Miners' Union Hospital in Wallace, where our sick and injured will be cared for in the future." The petition also contained an offer to maintain a local dispensary near the Bunker Hill mine.[31] According to union secretary James Drennin, the sole purpose of this proposition was to obtain "proper care and treatment of sick and injured miners, something we never received from company doctors." Clement rejected this proposal in a letter dated July 9.[32]

Shortly thereafter, Clement made a counterproposal for a company-supported hospital near Wardner, the control of which was described in nebulous terms. On the morning of August 6, the union replied with a modification of its position, advocating a strictly voluntary arrangement under which each employee could decide for himself whether to have one dollar deducted and contributed to the union hospital or to have nothing at all taken from his wages.[33] Later the same day, the company held a referendum on this issue. Three alternatives, none of which embodied either of the union proposals, appeared on the ballot. These were the choices: 1) continuation of the company doctor plan, 2) establishment of a company hospital, and 3) cessation of payroll deductions for health care, contingent upon "signing a contract with the Bunker Hill and Sullivan Mining Co., releasing them from all liabilities for sickness or injury while in their employ."[34] What the Wardner union president called a "fake election" was boycotted by the underground workers, although the nonunion workers in the ore-processing mill did participate. That evening, management posted a notice announcing the results in favor of the company hospital option and declaring that henceforth one dollar per month would be taken from every employee to support this scheme. The notice concluded by urging any workers who objected to this arrangement "to call at the company's office for their time."[35]

The union forcefully answered this challenge. Not long after the posting of the ultimatum, a spontaneous walkout by all but a handful of underground workers occurred on the night shift. This led to an impromptu union meeting at which a strike was declared. After the adjournment of the meeting, a large crowd of workers returned to the mine and induced the remaining four or five workers to join them. Three hundred miners marched from the mine in an orderly procession, leaving it completely shut down.[36]

The next day a union delegation met with management and put forward two demands. First, they reiterated their proposal to have deductions paid to the miners' hospital or to have no deductions at all. Second, the union called for a fifty cent raise for carmen and muckers, so that all underground workers would be receiving $3.50 per day. On August 8, local management received a telegram from the firm's newly appointed president, John Hays Hammond, which adamantly dismissed the union demands as "unjust" and "preposterous."[37] During the preceding year, Hammond, a prominent mining engineer and entrepreneur, had put together a powerful syndicate which was in the process of buying out Simeon Reed. On the same day the *Wardner News* published an appeal to the public by union secretary Drennin which asked the community to "support a union composed of men who act as their motto dictates: 'Act justly and fear not.' " Given that the Bunker Hill was the largest corporation in the district and that the new group of investors now in command included such formidable capitalists as Chicago reaper magnate Cyrus McCormick II, San Francisco banker William Crocker, and New York financier Darius Mills, the Wardner miners surely had plenty to fear.[38]

Following the initial walkout, the administration of the strike was relatively uneventful—and totally effective. At the request of Bunker Hill management, Sheriff R. A. Cunningham appointed thirty special deputies to protect company property at the firm's expense. But the sheriff found the strikers to be conducting themselves in a "quiet and orderly manner" and therefore refused to deploy the deputies after he had received a pledge from union leaders that no damage would be done. On August 10 the strikers were given an important expression of solidarity when the workers at the nearby Granite mine went out on a sympathy strike, vowing that they would not return until the Bunker Hill dispute was settled.[39]

On August 14, the company announced that it had seventy-five miners ready to return to work, including some "discontented strikers"; that it was advertising for more; and that it would therefore immediately resume operations. This strike-breaking tactic was either a bluff or a fiasco—a newspaper report the next day stated that "the Bunker Hill and Sullivan strike remains unchanged and no men are working in the mine." Union president Lew Roberts did, however, use the occasion of the proposed back-to-work movement to point out that "the scab who enlists on the side of capital against the just demands of his brothers is despised of all honest men and held in contempt even by those who hire him."[40]

Throughout the course of the strike the union attempted not only to discourage potential strikebreakers but also to recruit striking nonunion miners into their organization. By August 12, Roberts claimed that "with the exception of eight or ten all men working underground were union men, and the exceptions had promised to join."[41] Whereas rational means of persuasion were used with recalcitrant local miners, outsiders drawn to the district by Bunker Hill's inducements were forcibly deported:

> The Miners' Union of this place [Wardner] held an impromptu meeting last night and conducted two non-union men to the junction in a manner that indicated in no uncertain tone that the union men will not permit cut-rate artists to locate here. The procession presented a novel appearance as it moved down Main street with the scabs on a rail. The newcomers had offered their services for $2 per day. The sympathy of the citizens was with the union men in this instance.[42]

By August 20, Hammond complained that the strikers had extended their demands to include the firing of all foremen who were working during the dispute, an indication of who had the momentum at this point.[43]

Negotiations on August 21 produced a settlement of the strike. The agreement included an optional system of deductions under which employees were free to choose either the company doctor plan or no deduction at all. This represented a limited but significant concession to the workers, in that they had freed themselves from mandatory deductions. On the wage issue, Bunker Hill granted $3.50 per day for all underground workers, a complete victory for the union. The company agreed to rehire strikers without discrimination.[44]

At a meeting between labor and management on September 24, the terms of this settlement, which had appeared to be set, were changed, for reasons which are unfortunately unknown. The revised agreement granted further concessions: "Beginning October 1 next, this company will collect from every employee who has worked one day, one dollar per month, to be paid to the Miners' Union Hospital fund." In return for this capitulation, the union accepted the obligation to "furnish a resident physician and to assume all responsibility of furnishing necessary medical attendance to anyone injured or becoming ill while contributing to this fund."[45] The Wardner miners had decisively won the hospital battle of 1891. Thus assured of broad financial support, the unions proceeded to erect and transfer to the Sisters of Providence a sixty-bed facility, which served as a community hospital until 1967.[46]

In large part, the miners prevailed because their mutual-assistance programs fostered not only greater solidarity within their own ranks but also deep sympathy in the community. During the strike the union called attention to its charitable work and appealed for public support, consciously covering itself with a mantle of respectability. For example, on the day on which the company planned to resume work with strikebreakers, the *Spokane Review* published an appeal "To a Generous and Justice Loving Public" by union president Roberts. After reviewing the history of the unions' decision to build "a first-class miners' union hospital" and their willingness to allow a mine operator to sit on the board of trustees, Roberts asked that "all friends of labor . . . assist us to win this fight by notifying those in search of work to keep away from Wardner until the strike is won."[47] A few days later the same newspaper printed a lengthy statement by the Executive Board of the Coeur d'Alene Miners' Union which contended that the unions' self-help endeavors saved the taxpayers money: "We can say that our members suffering from sickness or injury and pecuniarily embarrassed are not thrown upon the public or [left to] die by the wayside as heretofore."[48] The miners also attacked the Bunker Hill corporation as a destructive and potentially tyrannical force in the community, directed by absentees in far-off San Francisco. Union denunciations of the company's refusal to allow its employees to spend their money where they chose undoubtedly struck a responsive chord with small businessmen threatened by company stores and boarding houses.[49] That these arguments succeeded is indicated in the *Wallace Press*'s observation that "the sympathy of the public is quite generally with the miners."[50] In such a favorable climate of opinion, strikebreakers could not be recruited locally, potential strikebreakers from outside the area could be run out of town with impunity, and the county sheriff refused to use his deputies.

Union benevolence also helped buttress demands for equal pay, explaining in part why this issue was interjected into the Bunker Hill strike after it was under way. The establishment of the hospital, together with the numerous union-conducted funerals, served to draw public attention to the extraordinary hazards shared by mining craftsmen and laborers alike. Beginning with the strikes at the Granite and Custer mines in July 1891, the miners maintained that "the shovelers have as hard and dangerous work underground as the miners and are entitled to the same pay." More specifically, union supporters argued that the introduction of "labor-saving machinery" had created equivalent levels of risk for miners and mine laborers, which justified uniform pay.[51] The work of the miners' union hospital produced irrefutable evidence of occupational injuries and illnesses among mine laborers. For example, on August 8 (the day after the Bunker Hill miners demanded equal pay), the *Wallace Press* published the first quarterly report of the hospital, which named all individuals treated and identified many of the conditions involved. These included "injured tibia done by falling earth and stones," "hurt by falling thro [sic] stope," "rheumatism with lead poisoning," and "bruised legs by ore car running on him."[52] A newspaper editorial advocating equal pay pointed to the lead poisoning problem brought to light by the hospital:

> A shovel full of lead-silver ore will weigh three and [sic] four times as much as a shovel full of white quartz, and to inhale the lead dust is the very essence of poison. Of course none of our mine owners have ever been "leaded," except politically, but the hospital records will show that sickness from that cause is quite prevalent. There is perhaps no harder task alloted [sic] to man than to shovel lead-silver ore for ten hours. We believe in grading employment [sic] about a silver-lead mine, but the underground workers should receive a uniform rate. . . .[53]

Throughout the recurrent skirmishes over this issue in the 1890s, the Coeur d'Alene unions relied heavily on the argument that equal risk demanded equal reward.[54]

Beneath its gleaming surface, the Gilded Age was also the "leaded" age, in which an evolving disease-intensive technology posed a multitude of health risks for American workers. In this period, for many reasons it was impossible for workers to force their employers to bear the health costs of production to any significant extent.[55] In the case of the Coeur d'Alene miners, the prevailing balance of forces was such that the miners could not have hoped to achieve more than the stripping away of the ideological veneer of corporate benevolence which surrounded employee health services and the supplanting of the company doctor with their own health care programs. The model of self-help fashioned in the Coeur d'Alene contributed not only to the subsequent establishment of more than twenty other metal miners' hospitals, but also to the founding of several hospitals in the Rocky Mountain region by locals of the United Mine Workers of America and the founding of the Union Labor Hospital in Eureka, California.[56] By illuminating the human cost of industrialization under capitalism and by demonstrating that the worker was able "to be his own benefactor," these institutions called into question the sufficiency and motives of employer efforts to safeguard the health of employees. Thus, mutual assistance was an important early stage in the evolution of the working-class movement to control health and safety hazards in the workplace.

NOTES

1. U. S. Bureau of the Census, *Benevolent Institutions,* 1904 (Washington: GPO, 1905), pp. 172–73; *Miners' Magazine* (June 10, 1909):4; *Miners' Magazine* (June 5, 1913):4; *Miners' Magazine* (May 1901):32–33; Silver City (Idaho) Miners' Union, Minute Book, vol. 1, pp. 164ff, Silver City Union No. 66 Records, Bancroft Library, University of California, Berkeley; R. L. Polk and Co., *Polk's Medical Register and Directory of North America,* 9th ed. (Detroit: R. L. Polk and Co., 1905), p. 1263; *Great Falls Daily Tribune* (Montana), March 27, 1902, p. 3; *Great Falls Daily Tribune,* March 29, 1909, p. 5; *Miners' Magazine* (Feb. 14, 1907):2; *Rhyolite Herald* (Nevada), Sept. 15, 1905; *Rhyolite Herald* April 26, 1907; Lucile R. Berg, "A History of the Tonopah Area and Adjacent Region of Central Nevada" (unpublished master's thesis, University of Nevada, Reno, 1942), p. 158; *Seven Troughs Miner* (Vernon, Nevada), Jan. 4, 1980; W. Edwin Newcombe (Physician, Lardeau Miners' Union Hospital, Ferguson, B.C.) to A. Shilland, April 29, 1903, Box 154, Folder 1, Mine-Mill Papers, Special Collections Division, University of British Columbia Library. All of these hospitals predated the founding of the ILGWU's Union Health Center in New York City in 1913, which is commonly considered the first labor health service facility.

2. U.S. Bureau of the Census, *Historical Statistics of the United States,* Part 1. Washington: GPO, 1975), pp. 606, 604; Rodman W. Paul, *Mining Frontiers of the Far West, 1848–1880* (New York: Holt, Rinehart and Winston, 1963), p. 149; U.S. Census Office, *Report on Mineral Industries in the United States at the Eleventh Census: 1890* (Washington: GPO, 1892); Robert W. Smith, *The Coeur d'Alene Mining War of 1892* (Corvallis: Oregon State University Press, 1961), p. 8.

3. Smith, p. 4; John Hays Hammond, *The Autobiography of John Hays Hammond,* vol. 1 (New York: Farrar and Rinehart, 1935), p. 181; U.S. Census Office, *Report on Mineral Industries,* pp. 80–81; John M. Henderson, William S. Shiach, and Harry B. Averill, *An Illustrated History of North Idaho* (Western Historical Publishing Co., 1903), pp. 996–1000, 1052–57; William S. Greever, *The Bonanza West: The Story of the Western Mining Rushes, 1848–1900* (Norman: University of Oklahoma Press, 1963), p. 276.

4. Thomas A. Rickard, *The Bunker Hill Enterprise* (San Francisco: Mining and Scientific Press, 1921), p. 75; D. E. Livingston-Little, "The Bunker Hill and Sullivan: North Idaho's Mining Development from 1885 to 1900," *Idaho Yesterdays* 7, no. 1 (Spring 1963): 36–39; Vernon H. Jensen, *Heritage of Conflict: Labor Relations in the Nonferrous Metals Industry up to 1930* (Ithaca: Cornell University Press, 1950), p. 27; *Engineering and Mining Journal* (June 6, 1891):663.

5. *Wardner News* (Idaho), June 1, 1888, reprinted in *Mining and Scientific Press* (June 23, 1888):397.

6. Smith, pp. 6–12; Job Harriman, *The Class War in Idaho,* 3rd ed. (New York: Socialist Labor Party, 1900), p. 3.

7. U.S. Census Office, *Report on Population of the United States at the Eleventh Census: 1890,* vol. 1 (Washington: GPO, 1895), pp. 494, 618; U. S. Senate, *Coeur d'Alene Mining Troubles,* 56th Cong., 1st sess., Document No. 24 (Washington: GPO, 1899), p. 13; *Engineering and Mining Journal* (March 3, 1894): 193; U.S. House of Representatives, Committee on Military Affairs, *Coeur d'Alene Labor Troubles,* 56th Cong., 1st sess., Report No. 1999 (Washington: GPO, 1900), p. 27; John Fahey, *The Days of the Hercules* (Moscow: University Press of Idaho, 1978), p. 182; A. C. Todd, "Cousin Jack in Idaho," *Idaho Yesterdays* 8, no. 4 (Winter 1964–65): 2–11.

8. Frank A. Crampton, *Deep Enough: A Working Stiff in the Western Mine Camps* (Norman: University of Oklahoma Press, 1982, 1956), pp. 22 (quotation), 22–25; Otis E. Young, Jr., *Black Powder and Hand Steel: Miners and Machines on the Old Western Frontier* (Norman: University of Oklahoma Press, 1970), pp. 30–40; Young, *Western Mining* (Norman: University of Oklahoma Press, 1970), pp. 178–91; Ronald C. Brown, *Hard-Rock Miners: The Intermountain West, 1860–1920* (College Station: Texas A and M University Press, 1979), pp. 68–73. The most helpful study is David Montgomery, "Workers' Control of Machine Production in the Nineteenth Century," in *Workers' Control in America* (New York: Cambridge University Press, 1979), pp. 9–31.

9. Mark Wyman, *Hard Rock Epic: Western Miners and the Industrial Revolution, 1860–1910* (Berkeley: University of California Press, 1979), pp. 84–117 and *passim;* Young, *Western Mining,* pp. 204–17; Paul, pp. 67–68.

10. Young, *Western Mining,* pp. 160–64; Young, *Black Powder and Hand Steel,* pp. 7–15; Richard E. Lingenfelter, *The Hardrock Miners: A History of the Mining Labor Movement in the American West, 1863–1893* (Berkeley: University of California Press, 1974), pp. 19, 33–34, 134, 162ff; Livingston-Little, p. 38; Harriman, p. 3; *Wallace Press* (Idaho), April 4, 1891, p. 1; *Spokane Review,* April 25, 1891, p. 3; William D. Haywood, *Bill Haywood's Book: The Autobiography of William D. Haywood* (New York: International Publishers, 1919), p. 80; Fahey, p. 23; *Wallace Press,* June 11, 1892, p. 1.

11. Lingenfelter, pp. 33–65, 131–34, and *passim;* Edward Boyce (charter member of the Wardner Miners' Union and president of the Western Federation of Miners), "Crime of the Century . . . ," in the U. S. Senate, *Coeur d'Alene Mining Troubles,* 56th Cong., 1st sess., Document No. 25 (Washington: GPO, 1899), pp. 1–2; Stanley S. Phipps, "From Bull Pen to Bargaining Table: The Tumultuous Struggle of the Coeur d'Alene Miners for the Right to Organize, 1887–1942" (unpublished Ph.D. dissertation, University of Idaho, 1983), pp. 10–14; Smith, pp. 20–21; *Wallace Press,* July 25, 1891, p. 4; *Wallace Press,* Aug. 8, 1891, p. 1; *Wardner News,* Aug. 8, 1891, p. 1.

12. Central Executive Miners' Union of Coeur d'Alene, *Constitution, By-Laws, Order of Business and Rules of Order,* revised (Wallace: Idaho State Tribune Print, 1895); Virginia City Miners' Union, *Constitution, By-Laws, Order of Business and Rules of Order* (Virginia City, NV: Brown and Mahanny, 1897), p. 3, quoted in Lingenfelter, p. 49; Lingenfelter, pp. 27, 227; Jensen, p. 10; Wyman, p. 179; James T. Smith (Deputy State Labor Commissioner of Colorado), "Testimony," July 13, 1899, in U.S. Industrial Commission, *Report,* vol. 12; *Relations and Conditions of Capital and Labor Employed in the Mining Industry* (Washington: GPO, 1901), p. 211.

13. James T. Smith, p. 14; Jensen, p. 249.

14. *Spokane Review,* Aug. 20, 1891, p. 5; Central Executive Miners' Union of Coeur d'Alene, "By-Laws [for affiliated local unions]," Article V, pp. 21–22.

15. *Spokane Review,* Aug. 20, 1891, p. 5; *Wallace Press,* April 25, 1891, p. 1; Central Executive Miners' Union of Coeur d'Alene, "By-Laws," Article VI, pp. 22–23; *Wardner News,* Feb. 28, 1891, p. 1; *Wallace Press,* May 30, 1891, p. 1; *Wallace Press,* Aug. 22, 1891, p. 1.

16. Charles Puckett, "Striking at the Bayonet Point," *Miners' Magazine* (April 30, 1908):14.

17. *Wallace Press,* Jan. 10, 1891, p. 4.

18. Lingenfelter, pp. 709, 23–26; Wyman, pp. 90–117; Brown, pp. 75–98; Larry T. Lankton, "The Machine *under* the Garden: Rock Drills Arrive at the Lake Superior Copper Mines, 1868–1883," *Technology and Culture* 14, no. 1 (Jan. 1983): 29–31; *Wallace Press,* June 11, 1892, p. 1: *Wallace Press,* May 28, 1892, p. 2;

Harriman, p. 3; U.S. Bureau of Mines, *Silicosis among Miners,* by R. R. Sayers, Technical Paper 372 (Washington: GPO, 1925), *passim;* Center for Law and Social Policy, *Miner's Manual: A Complete Guide to Health and Safety Protection on the Job* (Washington: Crossroads Press, 1981), *passim.*

19. *Coeur d'Alene Sun* (Murray, Idaho), March 9, 1889, p. 1: Henderson, Shiach, and Averill, p. 999.

20. *Wallace Press,* Sept. 21, 1890, p. 1; *Wallace Press,* Oct. 25, 1890, p. 1 (two articles).

21. *Spokane Review,* March 31, 1892, p. 2.

22. *Spokane Review,* Aug. 20, 1891, p. 5; Harriman, p. 3.

23. *Spokane Review,* July 16, 1892, p. 2; Harriman, p. 9.

24. Thomas A. Hickey, *The Story of the Bullpen at Wardner, Idaho* (New York: Socialist Labor Party, 1900), p. 4; Edward Boyce, p. 1; *Spokane Review,* March 27, 1892, p. 5; *Spokane Review,* March 31, 1892, p. 2; *Spokane Review,* May 13, 1892, p. 1; James R. Sovereign, "Testimony," in U. S. Industrial Commission, *Report,* p. 407; Harriman, p. 11.

25. Sovereign, p. 410.

26. Ibid., p. 407; "Pioneer," letter to editor, *Wallace Press,* June 11, 1892, p. 1.

27. *Wardner News,* March 7, 1891, p. 1; *Wallace Press,* April 4, 1891, p. 4; *Wallace Press,* April 11, 1891, p. 1.

28. *Wallace Press,* May 9, 1891, p. 1; *Wallace Press,* Jan. 16, 1892, p. 1; *Wallace Press,* April 11, 1891, p. 1; *Wallace Press,* April 4, 1891, p. 1.

29. *Spokane Review,* April 24, 1891, p. 8; *Spokane Review,* April 29, 1891, p. 5; *Wallace Press,* May 16, 1891, p. 4; *Spokane Review,* Aug. 12, 1891, p. 1.

30. *Wallace Press,* May 16, 1891, p. 4; *Wallace Press,* Aug. 8, 1891, p. 1; *Wallace Press,* June 13, 1891, p. 1; Mary McKernan, "Mother Joseph: Pioneer Nun," *American West* 18, no. 5 (Sept.–Oct. 1981): 20–21; Ellis Luria, *Seattle's Sisters of Providence* (Seattle: Providence Medical Center, 1978), *passim.*

31. *Wardner News,* Aug. 8, 1891, p. 1; *Spokane Review,* March 31, 1892, p. 2.

32. *Wardner News,* Aug. 8, 1891, p. 1.

33. *Spokane Review,* Aug. 9, 1891; *Wardner News,* Aug. 8, 1891, p. 1.

34. *Wallace Press,* Aug. 8, 1891, p. 4.

35. *Spokane Review,* Aug. 12, 1891, p. 1; *Wallace Press,* Aug. 8, 1891, p. 4; *Wardner News,* Aug. 8, 1891, p. 1.

36. *Wallace Press,* Aug. 8, 1891, p. 4; *Wardner News,* Aug. 8, 1891, p. 1: *Spokane Review,* Aug. 9, 1891, p. 1; *Spokane Review,* Aug. 12, 1891, p. 1.

37. *Wardner News,* Aug. 8, 1891, p. 1.

38. Ibid.; Hammond, vol. 1, pp. 186–87; Rickard, p. 77; *Spokane Review,* April 16, 1891, p. 3.

39. *Spokane Review,* Aug. 12, 1891, p. 1; *Spokane Review,* Aug. 13, 1891, p. 3.

40. *Spokane Review,* Aug. 14, 1891, p. 1; *Wallace Press,* Aug. 15, 1891, p. 1.

41. *Spokane Review,* Aug. 14, 1891, p. 1; *Spokane Review,* Aug. 9, 1891, p. 1.

42. *Wallace Press,* Aug. 22, 1891, p. 2.

43. *Spokane Review,* Aug. 21, 1891, p. 3.

44. *Spokane Review,* Aug. 11, 1891, p. 1.

45. *Spokane Review,* Sept. 26, 1891, p. 2.

46. *Wallace Press,* June 13, 1891, p. 1; *Wallace Press,* Feb. 20, 1892, p. 3; R. L. Polk and Co., *Polk's Medical Register and Directory of the United States and Canada,* 7th ed. (Detroit: R. L. Polk and Co., 1902), p. 481. By 1893 the facility had been renamed Providence Hospital. However, the unions continued to bear financial responsibility for the institution at least until 1905, at which time the original construction cost of $50,000 had been paid in full. See H. E. Bonebrake, Paul M.

Ellis, and E. J. Fitzgerald, *Medical History of Wallace Area* (n. p., n. d.): Sister Anselma Mary Price (Archivist, Sisters of Providence), personal communication, Jan. 28, 1982.

47. *Spokane Review,* Aug. 14, 1891, p. 1.

48. *Spokane Review,* Aug. 20, 1891, p. 5.

49. Ibid.; *Spokane Review,* Aug. 14, 1891, p. 1; *Spokane Review,* Aug. 12, 1891, p. 1. That many corporations in this period met similar opposition in small towns is shown in Herbert G. Gutman, "The Worker's Search for Power: Labor in the Gilded Age," in *The Gilded Age: A Reappraisal,* ed. H. Wayne Morgan (Syracuse: Syracuse University Press, 1963), pp. 38–68.

50. *Wallace Press,* Aug. 15, 1891, p. 4. During a strike in December 1890 at Burke, this newspaper proclaimed that "miners' unions are practically beneficiary [sic] societies, and it is an honor for all competent miners to belong to them." See *Wallace Press,* Dec. 20, 1890, p. 1.

51. *Wallace Press,* July 25, 1891, p. 1; *Wallace Press,* June 11, 1892, p. 1.

52. *Wallace Press,* Aug. 8, 1891, p. 1.

53. *Wallace Press,* May 28, 1892, p. 2.

54. *Spokane Review,* March 27, 1892, p. 5; *Coeur d'Alene Miner* (Wallace), Oct. 7, 1893, p. 4.

55. Carl Gersuny, *Work Hazards and Industrial Conflict* (Hanover, NH: University Press of New England, 1981), pp. 20–97; Gordon M. Bakken, "The Development of the Law of Tort in Frontier California, 1850–1890," *Southern California Quarterly* 60, no. 4 (Winter 1978): 405–19.

56. U.S. Bureau of the Census, *Benevolent Institutions,* 1910 (Washington: GPO, 1913), pp. 364–65; R. L. Polk and Co., *Polk's Medical Register and Directory of North America,* 13th ed. (Detroit: R. L. Polk and Co., 1914), pp. 1739, 1749; Daniel A. Cornford, "Lumber, Labor and Community in Humboldt County, California, 1850–1920" (unpublished Ph.D. dissertation, University of California, Santa Barbara, 1983), p. 445.

CHAPTER **2**   *Robert Asher*

The Limits of Big Business Paternalism: Relief for Injured Workers in the Years before Workmen's Compensation

The rise of capitalism in many European nations was accompanied by the erosion of paternalistic legislation and by a decline in paternalistic social practices by many employers of labor. "Free labor" in the "free market system" was often left to its own resources to deal with all vicissitudes. Similar developments in the United States were described by Alexis de Tocqueville, who believed that the factory system was creating a sharp social and economic demarcation between employers and workers:

> The manufacturer asks nothing of the workman but his labor; the workman expects nothing from him but his wages. . . . The territorial aristocracy of former ages was either bound by law, or thought itself bound by usage, to come to the relief of its servingmen, and to succor their distress. But the manufacturing aristocracy of our age first impoverishes and debases the men who serve it, and then abandons them to be supported by the charity of the public.[1]

Tocqueville's paradigm clearly is too extreme. But it does capture the direction of the legal and behavioral transformations that were impinging on the nation's nascent industrial labor force. Injured industrial workers, for example, could testify to the "freedom" that the judge-made law of industrial torts gave them—freedom to bear the costs of industrial hazards because employers were relieved of responsibility for negligent human action and negligent policies. And employers clearly had no legal obligation directly to care for free workers who were not serfs or chattel.

This is not to suggest that many manufacturers did not have altruistic impulses when they tendered aid to injured industrial workers. But this kind of benevolence was arbitrary: it was not given to all workers and was a charitable dispensation, not a legally mandated "right." In the United States between 1910 and 1917, all the major industrial states would enact workmen's compensation laws based on the principle that injured workers deserved relief. But well before the compensation era began, some companies, especially large corporations, instituted schemes (generally financed by worker and employer contributions) that regularized the granting of financial assistance to injured workers. These company relief plans were hailed by contemporaries as evidence of the humaneness of the men who managed the emerging industrial system and as an indication that the chasm between workers and employers would not lead to violent class conflict. In 1899,

Nicholas Paine Gilman concluded his study of contemporary corporate welfare programs, *A Dividend to Labor: A Study of Employers' Welfare Institutions*, with the observation that the dividends corporate paternalism
awarded to labor were

> one of Burke's "healing measures." . . . The employer is "made of social earth" as
> well as his operatives; and welfare institutions for their benefit, undertaken
> merely from long-sighted prudence, can hardly fail to bring him nearer as time
> goes by to a living sympathy with these men and women of like passions with
> himself.[2]

This chapter will examine the character of the systems for relief of
injured workers that were established by the nation's largest railroad and
manufacturing companies. Such corporate paternalism offered needed relief. But business largesse all too frequently came with many strings
attached, strings that reflected the tension between humane impulses and
the drive to pursue instrumental, self-serving objectives.

Industrial accidents, the frequency of which (in the United States) probably peaked around 1910, stemmed from employer negligence in combating
work hazards, from employer pressure to increase the intensity of labor, and
from risks inherent in industrial technologies. There can be no doubt that
some accidents were caused by worker recklessness and carelessness, the
latter often a reflection of fatigue and anxiety about the problems of low
income, unstable employment, and other hardships. Nor is there any question that ethnic and racial bigotry often produced a special insensitivity on
the part of employers and fellow workers toward the safety of workers
belonging to despised ethnic and racial groups. Long hours of labor and
compulsory overtime, dictated by market competition and the drive for
higher earnings, also contributed to accidents. Many employers ignored
known safety equipment and safety procedures because of the money and
time involved. Many employers flouted safety legislation, fired workers who
complained about hazardous conditions, and spent considerable sums on
legislative lobbying in opposition to stricter safety statutes. Before the
advent of workmen's compensation, many employers refused to admit state
safety inspectors to their premises. Injured workers who dared to sue their
employers for negligence resulting in damage to flesh and blood also knew
that they risked being fired for their temerity.

These factors often created a great deal of anger among workers, who
attributed industrial accidents to the prejudice and greed of their
employers.[3] Worker reactions to corporate paternalism must be viewed in
the context of worker views about the sources of the hardships they suffered.
What might seem to be generous relief to the injured may have been
appreciated, but business policies aimed at preventing damage to the body
and spirit in the first place would have been preferable.

Industrial workers frequently took up collections for their seriously maimed and deceased peers.[4] Most of the nation's early trade unions, established in the last quarter of the nineteenth century, originated as fraternal societies that paid members benefits for unemployment, sickness, and funeral expenses. When workers in the nation's most hazardous industrial occupation, the railroad industry, took steps to found labor unions in the early 1870s, their new associations invariably established insurance plans. "Benevolence, Sobriety, Industry" was the motto on the union seal of the Brotherhood of Locomotive Firemen (BLF). The placement of "Benevolence" before the other leitmotifs indicated how important the union-operated insurance plan was, both to the welfare of railroaders and to the appeal of the union to railroad workers, thousands of whom were hired anew by the railroads each year. At its sixth annual convention in 1879 the BLF noted that the hardships of the railroad worker, especially when he was injured, were grievous and that it was for "the prevention of evils of this description" that "this Order has been founded." In 1880, Hank Lovely—a Danville, Illinois fireman—proudly wrote to the *Locomotive Firemen's Magazine,* "I defy any man to show a benefit of that amount [$1000 for death] given [by the BLF] for the insignificant sum of seven dollars a year . . ."[5]

Our first information about employer accident relief plans appears in the years immediately following the Civil War. The Cambria Iron Works, one of the nation's earliest large-scale steel producers, placed the fines assessed against workers who violated company rules in a fund used to pay relief to injured workers.[6] The earliest accident relief fund financed jointly by workers and their employers appears to have been the one established in 1867 in New York City's brewery industry. In 1877 the Calumet and Hecla company, which operated many iron mines in Michigan, established a relief plan for its workers, and in 1884 the Westinghouse Machine Company founded a Mutual Aid Society, providing one-third of the Society's payments to injured workers. In 1875 the Philadelphia and Reading Railroad responded to a bitter strike by coal miners in its northeastern Pennsylvania fields by donating $20,000 to an accident relief fund. The road promised to take care of any deficits after the fund's first year of operation. In 1877 the railroad established some type of accident insurance plan for its engineers; the specific regulations of both plans have not survived.[7]

In 1877 executives of the Chicago, Burlington and Quincy Railroad considered some welfare benefits for their elite employees, the engineers. Vice president William Ackerman proposed guaranteeing the engineers lifetime employment, sickness pay, and relief payments for the families of engineers killed in the line of duty. Ackerman expected that such assistance would relieve his line's engineers of "their present feeling of insecurity and restlessness." Following the 1877 railroad strikes, another CB&Q executive favored some type of insurance plan "to show the men that while we would not submit to their dictation we still have their best interests at heart." But no action was taken until 1888, when the CB&Q established a relief fund modeled after those created by other railroads.[8]

In 1880 the Baltimore and Ohio Railroad established a relief association for its workers. The company gave the association an initial contribution of $100,000 and pledged to pay its administrative costs, which were at least $15,000 annually. All injured workers received free medical treatment. Existing employees were required to join the association, but were given the option of withdrawing before thirty days had passed. New workers had no choice. The B & O's workers expressed so much opposition to this infringement on their freedom that in 1888 the Maryland legislature revoked the charter it had granted to the relief association, which had formally been incorporated as a legal entity separate from the railroad. The B&O immediately reconstituted the fund as the railroad's Relief Department. All newly hired workers had to join. Veteran workers were told that if they wanted to be considered for promotion, they should enroll too.[9]

In 1882 the Northern Pacific followed suit. Announcing its fund to employees, the Northern Pacific issued a circular that began by noting that labor-capital antagonism was unwise and always injurious. Labor and capital depended on each other and had mutual interests. Outside parties (i.e., trade unions) had no business interfering with this symbiotic relationship:

> A corporation can rightfully expect entire loyalty and devotion to its interests on the part of employes, who should recognize no allegiance to Trade Unions or Brotherhoods as superior to that which they owe to their employers.

Of course, the employer had obligations: capital should promote the comfort and the "intellectual and moral improvement of all classes of its operatives." The Northern Pacific Beneficial Association (NPBA) was being founded in this spirit. Management pledged to cover the insurance plan's administrative expenses. Employee contributions would finance the accident and sickness benefits (the latter to be paid when an employee became sick while on the job).

The rules written by management for the NPBA stipulated that any injured worker who sued the railroad for damages would be denied NPBA benefits and would be expelled from the association. This provision meant that an injured worker who sued management for alleged negligence resulting in pain and suffering as well as a loss of income would be denied benefits that were to be financed from his own contributions and would, by virtue of expulsion from the NPBA, also be denied medical care and sickness pay. This was a severe punishment. The chances of becoming sick while at work were about equal to the chances of injury. Dollar amounts of NPBA medical care and sickness benefits generally equaled those of accident benefits. Thus the scheme of the Northern Pacific was a carrot that could be turned into a stick instead of a source of sustenance, discouraging lawsuits by injured workers. By avoiding tort action the railroad often limited its own direct liability for economic losses sustained by injured workers and also avoided the example that successful tort action would set for other injured railroad workers.[10]

Shortly after the NPBA was established the Northern Pacific issued a second circular noting that the NPBA's benefits would

> be largely in excess of the benefits payable in voluntary Unions of Association, where no guarantees [of solvency] are given and where, if the funds become exhausted there can be no further payments without additional assessments upon members.

The functional character of corporate benevolence was certainly not covert in this instance.[11]

The Pennsylvania Railroad, the nation's largest road, established a relief plan in 1886. Management called the scheme the Pennsylvania Voluntary Relief Department. Despite the freedom of choice the fund's name implied, new employees were required to join and all workers who applied for promotions had to be members. The "Pensy" provided that fines assessed against its workers would be added to the account of the Relief Department. The road also promised to pay the operating expenses of the Department and to make up any deficiencies in its accounts.[12]

Compulsory benevolence was also the mode of relief chosen by the Philadelphia and Reading Railroad when labor militancy spurred it to provide aid to injured workers. In 1887 the road's statistician warned President Austin Corbin that the railroad was beset by

> differences constantly arising between the P&RRRC and its employes, originating mostly from foreign influences and resulting, so disastrously, to all concerned. . . .

A relief fund would be

> the most expedient way by which an alienation of the men from orders such as the "Knights of Labor" may be made effective, thereby establishing a closer relationship [between the road and its workers].[13]

When Corbin discussed the plans for a relief association with his subordinates in the summer of 1888, he suggested that existing workers not be required to join but that membership be compulsory for new employees. But it appears that the railroad's division superintendents were determined to use every means at their disposal, short of outright dismissal, to convince recalcitrant workers to join the Relief Association. The division superintendents competed with each other to get workers to join the relief association. Veteran employees in one division were told that because they had refused to join the Relief Association, their superintendent had ordered their local agent to forward their names to him.[14] Another division superintendent, concerned that the general manager might think him too lax in his efforts to get employees to join the Relief Association, spelled out his efforts in a defensive letter to his superior:

> I am watching my chances when promotions are made, they will find themselves out of a job, the reason for which will show that we wanted to promote some other man more worthy. This method has already changed the minds of a number of those who refused, and I might say, two or three of which, have been promoted to position, after "backing water."

The superintendent then further explained that his coercive techniques did not violate the letter of the railroad's stated policy to not fire already employed workers who did not join the Relief Association:

> . . . none are discharged for this, but we will find a way to get them out without discharging those who will persist in not joining.

He concluded by observing that many of those reluctant to join were influenced by the temporarily low total take-home earnings caused by "reductions in the coal business." He predicted that when the present four days a week employment schedule expanded to six days a week, the older workers would readily join.[15]

These efforts appear to have been dictated both by a desire to secure the actuarial benefits of a large membership in the Association and by the knowledge that, without compulsion, many workers preferred to remain independent of the relief fund. A survey of the employees of the Shipping Department of the Philadelphia and Reading showed that as of September 15, 1888, 42 workers indicated after having read the prospectus of the relief plan that they would want to sign up and 45 employees told their supervisors they would not be interested in the plan. All the clerks and operators of the department were opposed to the plan. Workers on the coal piers divided: foremen favored the plan 9–5, clerks split 7–7, as did gang bosses, engineers were in favor of the plan by a margin of 8–6, and conductors approved it 9–4. We do not have any other such surveys, but this small piece of information suggests considerable opposition to the plan, especially since workers may have been somewhat reluctant to inform supervisors directly of their opposition to the company's initiative.[16]

Management concern about the hostility of the railroad's highly valued veteran workers to compulsory membership in the fund is suggested by a November 1889 letter in which the Reading's president told the assistant to the general manager that "I would not press anything upon the old employees, where salaries are less than $40. . . ." Corbin's wording suggests that he was making an exception to the administrative pressure only for the lower paid employees. All these actions ran contrary to the allegedly voluntary approach to existing workers, who in reality were subjected to severe pressure to join the Relief Association.[17]

When pragmatic necessity intervened, compulsory membership requirements were suspended. On August 28, 1893 the agent in charge of the Philadelphia and Reading's coal-carrying steam ships told his superiors that he was not forcing his men to join the relief association:

> . . . both sailors and firemen exhibit a strong disinclination against membership and at times when men are scarce it is absolutely impossible to secure reliable help . . . with an insistence upon membership in this Association . . . A strict insistence upon membership at all times would inevitably result in delays to the fleet as changes in crews are frequently made after business hours and some times at day break and men must be obtained at short notice.[18]

Functional altruism was truly flexible, as well as deceitful, professing voluntarism and practicing coercion, but reverting to a policy of freedom when coercion would literally have prevented the business from functioning.

Railroad workers understood that joint labor-management relief funds were designed to wean workers away from union activities, were compulsory and offered workers a choice between higher company plan benefits without the right to sue for negligence and lower union benefits that left workers free to initiate tort action. In 1877 the *Engineers' Journal* charged that the Philadelphia and Reading's first company insurance plan had been started to fill the vacuum created when the company had forced its engineers to withdraw from the Brotherhood of Locomotive Engineers and therefore from its beneficial plan.[19] A fireman in Iowa told the state's bureau of labor statistics in 1903 that the "Burlington relief is commonly called 'voluntary relief'" but was "erroneously named."[20] This rank and file railroader was less eloquent than Eugene V. Debs, whose first major editorial upon assuming the editorship of the *Locomotive Firemen's Magazine* in 1886 was a blistering attack on the relief department of the Pennsylvania Railroad. Debs characterized a flyer announcing the Pensy's new relief fund as "the address of the spider to the fly, in astute prose." Pointing out that the brotherhood relief funds were voluntary and were available to unemployed as well as employed railroaders, Debs focused on the coercive character of the railroad company fund, asserting that it did "not read well alongside the Declaration of Independence." Debs cited the case of an engineer killed in the employ of the Pennsylvania Railroad. His widow had received $10,000 in a tort action against the railroad. Under the Pensy's new relief plan, she would have received $1500. To Debs the choice between freedom and coercion was clear:

> No self-respecting man will barter his birthright, his manhood, his liberty and independence for a mess of pottage. He will not consent that any man or set of men shall control his earnings or any part thereof.[21]

William John Pinkerton, a railroad switchman who published his auto-biography in 1904, considered the railroad benefit funds "distasteful and unjust arrangement[s] whereby the employees are forced to shoulder the liabilities of the employer." Pinkerton felt that many members of the public, who did not understand the fact that injured workers or their widows had to sign liability releases to secure the benefits they had largely funded with their own money, were inclined to give the railroads too much credit for being altruistic when their funds paid out benefits to the fallen worker:

> When a beneficiary complies with the degrading terms of the payment of a policy, the munificent corporation grudgingly inserts its flabby hand, red with the blood of a thousand hearts, into the coffers of the pension fund, the fund to which the dead man contributed all the heavy years he served . . . and hands the widow HER OWN MONEY, while the world applauds the generosity and fatherly care of the corporation.[22]

On the other hand it seems likely that some railroad workers were grateful for management contributions to the principal and administrative expenses of the relief funds. It might be argued that workers also knew that their liability releases did not always rob them of a chance to sue, since many accidents were not attributable to any specific human act of negligence. But to workers who understood the relationship of accidents to the long hours of service and cost-minimizing management policies that created work hazards, their employers' voluntary spending on relief to the injured may have seemed a poor substitute for policies that would have *avoided* fatigue, stress, and injury.

Publicly and privately expressed criticism appears to have influenced some railroad managements to modify the coercive features of their relief funds. In 1890, the Northern Pacific changed the NPBA's rules: members were no longer required to sign a liability release before they received accident benefits. In 1898 Congress passed the Erdman Act, which dealt principally with railroad strike arbitration but included several clauses es-tablishing new federal policies for railroad relief associations. Compulsory membership in these associations was banned. Compulsory liability releases were prohibited. The statute stipulated that when railroads made cash payments to the principal of a relief fund, the management proportion of the relief fund's payments to an injured worker would be deducted from any award a worker obtained in a tort action. These provisions reflected the strong criticism that railroad workers had made of the coercive benevolence practiced by so many railroads. Their inclusion in a law that attempted to avert strikes suggests that the framers of the legislation—United States Commissioner of Labor Carroll D. Wright and Attorney General Richard Olney—well understood the antagonism created by the practices that were prohibited.[23]

Evidence on compliance with this law is understandably spotty. But it is clear that many major carriers evaded the intent of the law. By 1907 the Reading Railroad was making a distinction between temporary employees and permanent workers, requiring membership in its Relief Association only for the latter. Other railroads proclaimed a veneer of voluntary membership while pursuing a *de facto* compulsory policy. Consider the 1905 letter written by Joseph N. Redfern, the head of the Chicago, Burlington and Quincy's Relief Department, to a Northern Pacific official who was reviewing the issue of forcing new employees to join the NPBA:

> In the matter of new men coming into the service, the Company feels that if it wants one man, and two men apply for the job, one of whom says that he is willing to become a member and have the opportunity of protecting himself and his family in the event of disability or death, and the other says that he does not care about the Relief Department protection, that he will blow his money in the way he wants and that if anything happens to him he will expect the company or his fellow employees or the public at large to support himself and his family: then the Company selects the man who will become a member, feeling he will make the best employe.[24]

By 1908 the Pennsylvania Railroad no longer explicitly required employees to join the Relief Department. But an investigation by Gertrude Beeks, the National Civic Federation's expert on corporate welfare policies, revealed the existence of a *de facto* membership requirement:[25]

> When a man applies for employment he is handed a blank upon which, among other things, is the question, "Will you join the relief association?" If he says, "No," the chances are he is informed there is no position open. The average man who applies for a position is in such need that he will do anything to get it and frequently does not realize what he is signing.

By 1908 429 corporations had created funds to pay workers who were temporarily disabled by workplace injuries. 418 corporate funds covered fatal accidents. Interestingly, only 54 accident funds guaranteed benefits to permanently disabled workers, who had a greater need for relief than temporarily disabled workers. Overall, the benefits paid by the railroad accident plans were far superior to the awards by the accident funds of manufacturing and mining companies. Unfortunately, the U.S. Commissioner of Labor's studies of these plans totally ignored the issue of compulsory membership and mandatory liability releases.[26]

By the first decade of the twentieth century, some of the emerging giant manufacturing corporations, especially those in which J. P. Morgan's banking firm exercised significant influence, were moving toward a more generous treatment of injured workers. In 1901 Andrew Carnegie created the

Carnegie Relief Fund, donating $4 million for payments to workers injured in the employ of the Carnegie Steel Company. Carnegie announced that the fund was

> not intended to be used as a substitute for what the company has been doing in such cases . . . it is intended to go still further . . . I make this first use of surplus wealth upon retiring from business as an acknowledgment of the deep debt which I owe to the workmen, who have contributed so greatly to my success.[27]

Steelworkers who remembered the way Andrew Carnegie had smashed their union in 1892 and had thereafter increased working hours (which clearly increased injuries), establishing the hated twelve hour day, and had also steadily cut the wages of skilled steel operatives, might have regarded with a tinge of bitterness Carnegie's acknowledgment of their "contribution" to his "success."[28]

In 1908 the International Harvester Company became the next giant American corporation to establish a wholly company-financed compensation plan. As early as 1902 a company attorney had designed a plan that included a liability release. But Stanley McCormick, the brother of Company President Cyrus H. McCormick II, vetoed the idea of a release; he thought the workers would regard the plan as a sham if it was included. In 1904 an internal audit revealed that because the managers of the McCormick plants had been giving injured workers free medical care and limited financial aid, the company could save itself $10,000 yearly with a mutual benefit plan, including a 2 percent employee contribution, if the workers signed a liability release. This is the only statistical study I have found that specifically indicated the savings employers could anticipate by combining benevolence to the injured with a mandatory waiver of the right to tort action. When the Harvester Corporation established its plan in 1908, the directors rejected the advice of the Law Department that a liability release be included and instead followed the recommendation of Gertrude Beeks that the waiver of liability be discarded to avoid the "undue notice" that would have been "given this feature by the press, especially at this time."[29]

Beeks was referring to the unfavorable publicity the International Harvester Corporation had been receiving after publication of an April 18, 1908, feature article in *Collier's Weekly* that accused the company of using deceit in attempting to persuade a worker who had lost an arm in an accident to sign a waiver of liability on the basis of a $50 cash settlement. In addition, the directors of International Harvester were concerned about the strong opposition to their newly formed oligopoly from proponents of a more competitive, smaller-scale economic system. As Table 2-1 indicates,[30] the Harvester plan, which included a 2 percent worker contribution (but also included sickness benefits) was very generous, far exceeding any other corporation's voluntary benevolence. In implementing the plan, the com-

Table 2-1   International Harvester Corporation Accident
Compensation Plans

	1908 (2 percent worker contribution)	1910 (company financed)
Death	2 years' wages	3 years' wages (for death within 16 weeks of accident), $1500 minimum, $4000 maximum; 2 years' wages (for death 16–52 weeks after accident), $3000 maximum
Temporary disability	50 percent wages for maximum of 52 weeks	25 percent wages for first 30 days,* 50 percent thereafter for 2 years. If totally disabled after 2 years, pensioned at 8 percent of annual wages. Minimum of $10 monthly.
Permanent disability	50 percent wages for maximum of 52 weeks	loss of hand or foot—1.5 years' wages; loss of eye—39 weeks' wages; loss of both eyes, both hands, both feet, one hand and one foot—4 years' wages, $2,000 minimum

*Workers were given an opportunity to pay a small sum to receive 50% wages for the first 30 days of disability. Workers earning under $50 monthly paid 6 cents per month, those earning $50–100 paid 8 cents and those earning more than $100 paid 10 cents.

pany hoped to gain "the public opinion value of the voluntary act." George W. Perkins, chairman of Harvester's finance committee and the House of Morgan's chief liaison with President Theodore Roosevelt, noted that the plan had also been launched "with a view to anticipating any legislation that might be enacted in this country and alleviating just a little the strain between labor and corporate management." Within two weeks 82 percent of the company's workers had joined up. Company superintendents reported that "without exception . . . the matter had been splendidly received by the men, that they feel the Company had done a generous thing and that it is appreciated." Many newspapers praised International Harvester as an enlightened employer. And President Theodore Roosevelt was favorably impressed.[31]

In 1910 the International Harvester Company increased the benefits of its compensation plan and eliminated worker contributions. And United States Steel, also controlled by the House of Morgan, announced its own compensation plan. The steel company feared antitrust action and was also concerned about the criticism of journalists and social investigators—

especially two volumes of the Pittsburgh Survey, John Fitch's *The Steelworkers* and Crystal Eastman's *Work Accidents and the Law*. Moreover, as the Corporation's chief attorney privately noted in 1909,

> The need for the consideration of this question was shown by the fact that in the legislatures of nearly a dozen of our states [workmen's compensation] . . . legislation has been introduced and is being urged more generally and more determinedly each year. . . . Already many of the common law defenses of the employer have been done away with in a large number of our states and the decision of most of the questions in accident litigation is left to the jury. This makes it more and more necessary to avoid litigation to the greatest extent possible by amicable settlements . . . I believe that such a plan of accident relief as is proposed . . . would do much to prevent the bringing of suits against our companies.[32]

Economic and political pressure, not altruism, were the dominant impulses governing the new benevolence U.S. Steel began offering to its workers.

In 1908, U.S. Steel began to invest large sums of time and money in a safety drive that produced a 43 percent decline in serious accidents by 1912. Ironically, the stock bonus plan the company had instituted for its higher paid workers in 1902 had led foremen and supervisors to increase the pace of work as they strove for higher productivity. This speed-up had led to an increase in accident frequencies and had created greater worker fatigue, which damaged worker health.[33]

Comparison of the two most generous corporate compensation plans in the United States in 1910 with the first state workmen's compensation laws indicates that the voluntary benevolence of welfare capitalists was not up to the standards of New York's 1910 law, a statute that provided lower benefits than subsequent pioneering state compensation statutes. In New York, families of workers killed in accidents were to receive 3 years' wages. U.S. Steel offered widows of men with less than 1.5 years of service only 1.5 years' wages. The company did make a provision for increasing the benefits, with 10 percent extra awarded for each child under 16 and 3 percent extra for each year of service over 5 years. Considering the high turnover rate in the industry, this 3 percent bonus for service did not help most workers. For a widow to receive death benefits equal to the New York state law's level, she would have had to have ten children under age 16. International Harvester's 1910 plan offered widows benefits equal to the New York statute.

Temporarily disabled workers received 50 percent of their wages in New York state, for a period of up to 8 years. International Harvester's 1910 plan paid workers only 25 percent of their wages for the first thirty days of disability and 50 percent for the next 2 years; thereafter, workers were to receive 24 percent of their wages. U.S. Steel only paid temporary disability benefits for one year and offered single men with less than 5 years of service only 35 percent of their wages. Workers with more than 5 years of service

were awarded 2 percent of wages for each additional year of service, with a
maximum benefit set at $9, which was $1 below the $10 maximum in New
York state.[34]

Historians of welfare capitalism have often argued that the benefits
American businesses extended to their workers helped to reduce the appeal
of trade unionism to working people.[35] Statistical verification of this hypoth-
esis is not possible. We can simply observe that workers in companies
with well-funded welfare systems were both quiescent and militant, depending on the level of employment demand in the industry, the tradition
of trade unionism in the industry, the severity of the direct repression
of union activity by employers, the degree of skill required, employee
unity along ethnic, racial, and gender lines, and a host of other factors. We
do know that railroad workers became *more* unionized as the nineteenth
century wore on. But we cannot scientifically test the assumption that in the
absence of employer paternalism labor militancy would have reached higher
levels.

Many railroad workers, as well as workers in other industries, must have
been alienated from management because of the particularly contradictory
character of corporate paternalism in the area of industrial safety and accident compensation. Company-instituted relief plans frequently combined
generosity with compulsion and favoritism toward veteran employees in a
manner that was obnoxious to many working people. Moreover, some workers interpreted the positive aspects of corporate paternalism as a defensive
response to worker organization. Hayes Robbins, who was active in the New
England Civic Federation's efforts to promote both collective bargaining and
welfarism, noted in 1906 that workers who wanted union recognition and
higher basic wage scales often interpreted corporate welfarism "as a substitute for what is denied them."[36] In the twentieth century, many American
workers understood that only threats of adverse publicity, antitrust action,
unionization, and political socialism made employers respond with programs
that increased the amount of aid extended to injured workers and improved
on-the-job safety. Even then, corporate paternalism was contradictory,
stressing safety and higher levels of accident compensation while striving for
a greater intensity of labor, which increased the likelihood of fatigue and
accidents.

The contradictions of welfare capitalism reflected the imperatives of a
competitive, profit-driven economic system. Corporate managers strove to
maximize their control over the labor process and to augment profits by
minimizing expenditures. Consequently, humane considerations were often
diluted or were combined with coercion in an attempt to reduce worker
challenges to management autonomy. Corporate paternalism directed
toward injured workers became less authoritarian and tendered greater
financial benefits as American workers formed stronger trade unions and
engaged in more vigorous and more radical political agitation.

NOTES

1. Alexis de Tocqueville, *Democracy in America*, vol. II (New York: The Colonial Press, 1899), pp. 196–97.

2. (Boston: Houghton, Mifflin and Company, 1899), pp. 352–53.

3. These factors are explored in more detail in Robert Asher, "Industrial Safety and Labor Relations in the United States, 1880–1915," *Life and Labor: Dimensions of American Working Class History,* ed. Charles Stephenson and Robert Asher (Albany: State University of New York Press, 1986); see also the cogent analysis of Kurt Ketzel, "Railroad Management's Response to Operating Employees Accidents, 1890–1913," *Labor History* 21 (Summer 1980): 351–68; and Charles Hugh Clarke, "The Railroad Safety Movement in the United States: Origins and Development, 1869 to 1893," (doctoral dissertation, University of Illinois, 1966).

4. For a record of one such collection, see the 1867 collection sheet in the papers of the Cleveland Cliffs Iron Company, Archives, State of Michigan, Lansing, MI.

5. *Locomotive Firemen's Magazine* 3 (August 1879): 3; 4 (October 1880): 311; 4 (November 1880): 333.

6. Robert Asher, "Workmen's Compensation in the United States, 1880–1935," (doctoral dissertation, University of Minnesota, 1971), p. 47.

7. Nuala M. Drescher, "The Workmen's Compensation and Pension Proposal in the Brewing Industry, 1910–1912: A Case Study in Conflicting Self-Interest," *Industrial and Labor Relations Review* 24 (October 1970): 36; Harold Aurand, *From the Molly McGuires to the United Mine Workers: The Social Ecology of an Industrial Union, 1868–1897.* (Philadelphia: Temple University Press, 1971), pp. 160–61.

8. Thomas C. Cochran, *Railroad Leaders, 1845–1890* (New York: Russell and Russell, 1965), pp. 178, 236.

9. William F. Willoughby, *Workingmen's Insurance* (New York: Thomas Y. Crowell & Company, 1898), pp. 285–86.

10. Circular to Employees of the Northern Pacific Railroad, 1880. Northern Pacific Beneficial Association (NPBA) records, Northern Pacific Railroad Company, St. Paul, MN.

11. Circular, Aug. 28, 1882, NPBA Scrapbook #1, NPBA records.

12. *Locomotive Firemen's Magazine* 10 (April 1886): 194.

13. Charles Hunter to Austin Corbin, Dec. 29, 1887, Philadelphia and Reading Railroad Company (P&RRRC) Papers, Historical Society of Pennsylvania.

14. G. M. Lawler, Superintendent, Williamsport Division, to A. A. McLeod, General Manager, March 12, 1889; and Charles Wilson to A. A. McLeod, Feb. 12, 1889, P&RRRC Papers.

15. Lawler to McLeod, P&RRRC Papers.

16. John O. Keim, Agent, to I. A. Sweigard, General Superintendent, Sept. 15, 1888, P&RRRC Papers.

17. Austin Corbin to A. H. O'Brien, Asst. to G.M., Nov. 25, 1889, P&RRRC Papers.

18. F. W. Taylor to James Landis, Chief Clerk, Aug. 28, 1893, P&RRRC Papers.

19. *Engineers' Journal* XI (May 1877): 219; (July 1877): 314.

20. Iowa, Bureau of Labor Statistics, *Annual Report,* 1903, p. 496.

21. *Locomotive Firemen's Magazine* 10 (April 1886): 193–200.

22. William John Pinkerton, *His Personal Record: Stories of Railroad Life* (n.p., 1904), pp. 246, 252–53.

23. Minutes of the NPBA Board of Managers, Book A, 119, NPBA records; *Congressional Record*, Fifty-Fifth Congress, Second Session, XXX, 4638, 4659; Gerald Eggert, *Railroad Labor Disputes: The Beginnings of a National Strike Policy* (Ann Arbor: University of Michigan Press, 1967), pp. 217–19.

24. A. Voorhees to W. C. Brisler, Claim Agent, June 24, 1907, P&RRRC Papers; Joseph Redfern to M. Dering, March 25, 1905, Northern Pacific Railroad Papers, 1.C.5.2F, Minnesota Historical Society.

25. Gertrude Beeks, "Memorandum on Pennsylvania Railroad, 1908," Cyrus Hall McCormick, Jr. Papers, Box 29, State Historical Society of Wisconsin.

26. U.S. Department of Commerce and Labor, *Twenty-Third Annual Report of the Commissioner of Labor,* 1908 (Washington, DC: Government Printing Office, 1909).

27. Crystal Eastman, *Work Accidents and the Law* (New York: Charities Publication Committee, 1910), pp. 161–62.

28. Ernest Cotnoir, "The Homestead Strike" (unpublished honors thesis, The University of Connecticut, 1983).

29. Report of Committee on Benefit Insurance, July 10, 1908, Box 18, George W. Perkins Papers, Columbia University; see also Cyrus H. McCormick, Jr. to Perkins, July 13, 1908, E. A. Bancroft to Cyrus H. McCormick, Sr., July 14, 1908, Memorandum of Conference between C. F. Bonney, Manager Factory Department, Continental Casualty Company and G. A. Ranney, July 13, 1908, Box 19, Perkins Papers.

30. Based on Crystal Eastmen, *Work Accidents,* 304–306; Robert Ozanne, *A Century of Labor-Management Relations at McCormick and International Harvester* (Madison: The University of Wisconsin Press, 1967), p. 79.

31. Mary L. Goss, "Benefit or Relief (Industrial) Association or Workingmen's Insurance: A History of the Movement," July 13, 1908, Box 19, Perkins papers; Cyrus H. McCormick, Jr. to Perkins, Sept. 4, 1908, Oct. 17, 1908, Box 19, Perkins Papers; Perkins to Edward C. Grenfell, Dec. 19, 1910, Box 21, Perkins Papers.

32. Charles Mac Veagh, General Solicitor, United States Steel Corporation, to William B. Dickson, Dec. 21, 1909, William B. Dickson Papers.

33. Charles A. Gulick, Jr., *Labor Policies of the United States Steel Corporation* (New York: Columbia University Press, 1924), pp. 141–70; John A. Garraty, *Right-Hand Man: The Life of George W. Perkins* (New York: Harper and Row, 1957), p. 114.

34. Chapter 674, *Laws of New York,* 1910; Chapter 467, *Laws of Minnesota,* 1913; Eastman, *Work Accidents,* pp. 300–303.

35. See especially David Brody, "The Rise and Decline of Welfare Capitalism," in *Change and Continuity in Twentieth Century America: The 1920s,* ed. John Braeman et al. (Columbus: Ohio State University Press, 1968), pp. 147–78.

36. Hayes Robbins to Gertrude Beeks, Feb. 19, 1906, National Civic Federation Papers, Box 107, New York Public Library.

CHAPTER **3** *Anthony Bale*

America's First Compensation Crisis: Conflict over the Value and Meaning of Workplace Injuries under the Employers' Liability System

During the first decade of the twentieth century a crisis developed over the manner in which injuries produced by work were compensated, leading to fundamental changes in the way injuries at work became transformed into monetary compensation. Class struggle over accidents in workplaces, in the courts, and in state legislatures led to a rise in the value of the legal rights of action for occupational injuries. The system for compensating workplace injuries served to inflame class antagonism and expose industry to reformers' accusations that industrial negligence bordered on murder and pushed deserving families into poverty.

This account extends the work of previous writers on the forces and the historical conjuncture leading to passage of the workers' compensation laws in the second decade of the twentieth century. James Weinstein showed the active involvement of corporate elites in the passage of the workers' compensation laws; however, Weinstein did not lay out the forms of class struggle that helped induce these elites to push for replacing the employers' liability system.[1] Robert Asher's dissertation is the single most complex and comprehensive account of the passage of the workers' compensation laws.[2] Asher analyzed the social forces responsible for the laws, and his extensive state studies produced a picture of some of the different configurations that resulted in different forms of this "conservative reform." Asher showed that "the movement for workmen's compensation was a coalition of tenuously affiliated social classes and organizations" that varied within diverse state political cultures.[3]

This chapter isolates deep structural elements of the crisis preceding passage, organizing them around a central notion of "compensation crisis" that is also useful in analyzing more recent similar situations, such as that involving asbestos litigation. I follow up on the insight of America's leading socialist student of social insurance, I. M. Rubinow, that the rising value of the injured worker's right of action against the employer was at the center of capital's interest in a new system.[4] This study documents both the rise in the value of the right of action and some of the elements of class struggle that contributed to this rise.

Class struggle over the compensatory value of the labor power expended by workers in the form of bodily suffering and death produced in the labor process is every bit as much class struggle as the more widely recognized forms undertaken to raise the value of labor power at the point of production. The fact that this compensatory class struggle occurs in a complex process mediated by courts, lawyers, legislatures, and most centrally by the insurance industry may account somewhat for the lack of attention to this phenomenon to date among those studying class struggle in America. This struggle was acted out largely through adjudication of individual cases, rather than through collective action.

An important element in this account is the idea that much of what goes on around workplace injuries in the courts, press, regulatory arena, shop floor, unions, etc. constitutes a discourse concerning the meaning and consequences of these events. This discourse concerning the cost paid in suffering by those who create commodities and wealth is inherently moral in nature. Elements of this discourse at times threaten to become a critical inquiry into the way capitalist production relations become manifest in violent, preventable bodily suffering. The discourse includes attributions and action involved in interpreting connections between bodily suffering and work; it includes establishing the practical consequences, such as who should bear what financial share, of injuries established as work-related. The discourse may involve reading manifest suffering as connected to work relations that are homicidal in character; it opens up a zone of contest as to who and what social relations are at fault.

Concealed in the commodities that capital creates are acts of homicide produced in the conditions which capital creates to produce those commodities. The discourse around workplace injuries and illnesses contains within itself the potential for a critique of capitalist relations of production as murderous in essence. The shift involved in moving from a fault-based inquiry, in the employers' liability system, to a no-fault workers' compensation system was a way of shortcutting some of the gains workers and their allies had made in this discourse by calling into question the moral legitimacy of capital.

Workers created a compensation crisis around the value of the lives and limbs lost at work. This crisis had two primary aspects and two secondary aspects leading to a conjuncture that resulted in the resolution of the crisis, in the 1910s, through passage of workers' compensation laws. First, the value of the right of action for work-related injuries and the uncertainty of future outlays rose sharply, reflecting victories in the courts and legislatures. Second, the employers' liability system inflamed antagonisms in class relations by providing small amounts of compensation through a process of lengthy litigation usually resulting in small settlements. The extensive discourse of fault conducted around workplace accidents resulted in a continuing examination of the pain and suffering inflicted on workers by capital in the process of production. Third, contributing to the crisis was the widespread nature of the accident problem: every region was impacted, and a substantial portion of the population was threatened. Last, the rising concern with compensation for work accidents occurred in the context of rising working-class radicalism and threats to the newly ascendant large corporations. Compensation of workplace accidents became the focal point of an effort by business to reduce class antagonism, by workers to raise the value of the lost vitality incurred at work, and by social reformers to create the first system of social insurance in the United States.

In the employers' liability system the value of injuries was set through negotiations between workers on the one hand and employers and their insurance companies on the other, in the shadow of a complex and changing set of legal rules. Most injuries were never compensated in this market; for most of the injuries that entered the market, a small settlement was negotiated; a small proportion of injuries were settled by trials in civil court, continually establishing the value of the other injuries. Judges controlled which cases could get to juries, which laws juries were to apply, and the appeals process through which jury verdicts could be overturned.

Under the employers' liability system, the employer was held negligent if he violated the common law duty of providing a reasonably safe place to work. However, employers had three famous defenses that could bar recovery even in the face of employer negligence: the fellow servant rule, contributary negligence, and assumption of risk. The fellow servant rule made recoveries from employers impossible when the injury was deemed to be caused by the negligence of a fellow worker. If the employee could be found to be at fault to any degree, a suit could be disallowed under the doctrine of contributory negligence. And workers were held to assume risks of which they were aware or could reasonably have known of even if they involved unsafe conditions.

These three defenses provided a strong legal shield, preventing most injured workers from obtaining even a day in court, as well as a powerful club in bargaining for settlements. They were part of the legal fiction of an employment contract in which the worker's presence on the job was held as largely creating a waiver of liability for whatever injury befell him. The three defenses—particularly the fellow servant rule—came under legislative and judicial attack in many states, although few major changes were made. Similarly, attempts were made to strengthen the employer's duties by such measures as creating presumptions that violation of a safety statute constituted negligence.

This chapter focuses on the creation of a compensation crisis for capital. Under the employers' liability system, workers were liable for almost all financial consequences of a work-related accident, subsidizing capital accumulation in the process. Through access to juries and through the creation of the discourse of fault calling into question the human suffering extracted in the process of production, workers and their reformer allies turned the employers' liability system itself into a compensation crisis for capital.

THE WIDESPREAD ACCIDENT PROBLEM

Death and injury from accidents of all types were common occurrences in America in the late nineteenth and early twentieth centuries. In 1908 and 1909, approximately 10.5 percent of deaths among all workers occurred from some form of accident.[5] Roger Lane notes that in the late nineteenth century the electrified street railways in Philadelphia "accounted for roughly as many fatal accidents, in proportion to population, as motor vehicles in the modern city."[6] Litigation over street railway accidents in the late nineteenth century was more extensive than that for workplace injuries; in 1900, 42 percent of all accidental injury lawsuits in Boston's courts were from street railway injuries, as opposed to 12 percent for industrial accidents.[7]

Electrified streetcars were just part of a new hazardous mechanical environment. Elevators, electricity, gas stoves, and gas systems were other elements that produced numerous serious injuries for people at work and away from it. The size, power, and speed of this environment helped create a new level of risk from accidents involving the new technology. Philadelphia's death rate from casualties—usually falls, work and transportation accidents—rose from 23.2 per 100,000 per year from 1860–1866 to 34.8 from 1881–1887, and to 39.5 from 1895–1901.[8] Burns and scalds added another 11.3 accidental deaths in the 1895–1901 period, roughly the same rate as in the 1860s.

The accident problem, of which industrial accidents were a large part, impacted on a large proportion of families, businesses, and towns. Workers and employers fought in numerous state legislatures over safety codes. Disasters helped keep the issue in the public eye. From 1905–1909, there were 85 major coal mine disasters, taking 2,640 lives.[9] Mine explosions in Monongah, West Virginia (1907), Jacob Creek, Pennsylvania (1908), and Marianne, Pennsylvania (1909) killed 358, 239, and 154 miners respectively.

Frederick Hoffman of the Prudential Life Insurance Company developed the most widely accepted estimates for industrial accidents in the early twentieth century. Hoffman estimated that in 1913 there were approximately 25,000 industrial fatalities and 700,000 injuries involving a disability of more than four weeks among the 38 million employed men and women.[10] Between 1907 and 1912, among all males, 9.4 percent of deaths were caused by accidents; among coal miners the proportion was 23.2 percent, among powder makers it was 72 percent, and among electrical lineman it was 49.6 percent.[11] Hoffman cited evidence that from 1898–1907, one out of every 230 railroad switchtenders died every year, as did one out of every 131 railway trainmen. People who did not experience high risk on the job might well encounter the risk of injury from the new industrial environment in other aspects of their lives and seek redress in courts.

RISING VALUE OF THE RIGHT OF
ACTION FOR OCCUPATIONAL INJURIES

At the heart of the compensation crisis was the expansion of the value of the right of action at law for work-related injuries. Workers developed greater access to juries with broadened grounds for winning their suits. Rather than suing small community-based enterprises, often the worker was suing a large company disliked by many of the pool of jurors. An active plaintiffs' bar, working on contingency fees, helped raise the value of the right. Large companies developed benefit plans and partial compensation schemes, most often for medical expenses, and attempted to arrange settlements before or after suits were started. Pulling this system along were large judgments won in court and the prospects of more and larger awards.

The juries hearing industrial injury cases were likely similar to those in turn-of-the-century Boston studied by Silverman: "These were, roughly speaking, juries of peers."[12] Many more prominent members of the community were either barred from jury duty because of their occupation or easily released from it. Unskilled workers were greatly underrepresented as both plaintiffs and jurors. The largest blocs of Boston civil jurors in 1900 came from lower level white collar occupations; fully 40 percent of Boston civil jurors in 1900 were skilled or semi-skilled blue-collar workers.

The state courts of last resort stood at the apex of the employers' liability system for compensating occupational injuries. By interpreting the law and reversing lower court verdicts for injured workers, the courts could control the value of the common law right to action for occupational injuries. The value of this right at the appellate level helped set the value at the point of the most common transactions farther down the system: employers offering no compensation whatever, small payments for medical bills and/or funeral expenses, or small out of court settlements for lost income.

A higher percentage of industrial injury verdicts for plaintiffs were reversed at the appellate level for industrial injuries than for most other types of negligence, reflecting the strength of the employers' defenses. In New York State before 1900, of the thirty industrial injury appellate cases where the verdict had been rendered for the plaintiff, 28 were reversed by the higher courts; of the three verdicts for defendants from the lower courts, all were upheld on appeal.[13] "In nineteenth-century California, the employee prevailed in thirty-seven Supreme Court opinions, the employer in forty-three; in the New Hampshire Supreme Court, there were eleven victories for employees, twenty-three for employers."[14] In Wisconsin in the 1890s, 72.8 percent of the appellate cases involving industrial injuries were decided for the worker in the lower courts; however, the Supreme Court decided only 35.1 percent in their favor.[15] But by 1905–1907 the odds had shifted a bit in favor of the workers: the lower courts in Wisconsin decided 62.5 percent of the appellate cases for the worker, while the Supreme Court decided 47.5 percent for them.

Robert Asher argues that 1905 was a watershed in the fortunes of workers in the courts:

> After 1905 judges in many higher state courts appear to have become more sympathetic to the hardships suffered by injured workers. With increasing frequency verdicts were allowed to stand and judicial reinterpretation of the common law increased the chances for injured workers to recover damages by suing their employers.[16]

Labor campaigns for employers' liability laws weakened some employer defenses and helped dramatize the plight of injured workers to the pool of already largely sympathetic potential jurors.[17] Workers felt employers' liability laws would be as much preventive measures as means of compensation. John Flora, a leader of the Chicago Federation of Labor, argued that workers "want to make it so expensive for the employers to kill their workmen that every safety appliance known to science will be installed."[18] Although the movement for repeal of the employers' defenses was largely unsuccessful, the small gains and continual pressure from the labor and reform forces meant that their further erosion was likely, or at least possible. By 1910, victories by labor in the legislatures of many industrial states had created a body of occupational safety and health regulation comparing favorably to that in Germany and England.[19] Safety and ventilation statutes made it easier to prove employer negligence in some states where the courts held that such violations constituted a showing of employer negligence; their passage was a sign of interest in the plight of injured workers—interest that carried over to jurors and judges.

As the 1900s progressed, workers appear to have had a better chance of keeping the awards won before juries. Illinois appellate court decisions between 1905 and 1910 were two to one in favor of injured workers.[20] "In 1908 and 1909 the Minnesota Supreme Court favored the injured plaintiff in seventy-three percent of its decisions, while the highest courts of Pennsylvania, Michigan, Wisconsin, and Iowa posted forty-three percent, forty-six percent, fifty-three percent and sixty-three percent records, respectively."[21] In Ohio from 1906–1908, 80 percent of the verdicts in nonfatal accident cases in the Common Pleas and United States Circuit Court were decided in favor of the injured worker.[22]

Premiums paid for employers' liability insurance rose rapidly, going from $203,132 in 1887 to $7,129,444 in 1900, rising to $15,767,818 in 1905, almost doubling in three years to $27,938,311 in 1908, and reaching $35 million in 1911.[23] Approximately two-thirds of businesses in the last half of the first decade of the twentieth century carried employers' liability insurance.[24] In Wisconsin in 1906 an additional seven percent of businesses had workmen's collective policies instead of employers' liability policies. Workmen's collective policies insured employees for bodily injuries at work; a typical policy might cover losses for specified injuries or total disability for a certain number of weeks' wages up to a maximum amount with benefits paid on a no-fault basis.[25] Nationally in 1906 premiums for workmen's collective insurance came to $465,000, approximately two percent of the premiums paid for employers' liability insurance in the same year.[26]

Rises in premiums over the long run reflected rises in defense costs and company expenses more than increased amounts going to injured workers. From 1898 to 1902, Massachusetts insurance companies writing employers' liability policies in all the states charged $38,200,000 in premiums and paid out 63 percent of the premiums in losses; from 1903–1907 these same Massachusetts insurers took in $74 million in premiums and paid out only 37 percent in losses.[27] Large increases in losses could be made up in following years by higher premiums. In 1906 employers' liability losses nationally rose by sixty percent and the percentage of premiums paid out to injured workers rose from 33 percent to 45 percent. By 1908 premiums had risen to keep pace so that only 36 percent of the premiums were paid out as losses.[28] In Minnesota the increase in losses was so rapid that premium rises couldn't keep pace. Between 1903 and 1907 Minnesota employers' liability insurers paid out 58 percent of the premiums; however, "in 1908 the figure jumped to 68 percent, and one year later went to an incredible 78 percent."[29]

Helping fuel the rising damages and premiums was the success of workers and attorneys in raising the value of the legal right of action for workplace injuries. Croyle has analyzed a sample of Minnesota accidents in 1909 and 1910. Holding other things equal, a worker in a nonfatal accident received $551 more compensation if he went to court than if he did not. For each dollar spent on a lawyer in nonfatal cases the worker realized an additional $2.61 in compensation; in fatal cases each dollar in legal fees brought the family $3.02.[30]

Uncertainty of the outlays for compensating industrial accidents became a growing problem for employers. In a study of work-related fatalities in Manhattan in 1908 and in Erie County, New York in 1907–1908, where a case had been closed, 41 percent received no compensation while 2.1 percent received over $5,000, with the highest award being $10,875. The 3 percent of the cases decided in trials averaged a recovery of over $5,000; those settled out of court averaged a recovery of $1,477, and those settled without a suit averaged $703.[31] Although the general pattern of a few large jury awards, many small settlements, and many getting funeral expenses or nothing at all was not changing substantially, the greater likelihood of big awards coupled with a higher value of settlements purchasing the worker's right of action were creating uncertainty as to present and future expenses for employers. An employer with a liability policy would typically be covered for $5,000 of a $10,000 verdict; liability insurance might not provide total protection from large losses. Jury determinations of the value of life and limb along with their distaste for large, negligent companies horrified employers. At the 1910 convention of the National Association of Manufacturers an employer related a story of a verdict the week before "of 41,000 dollars for loss of arms against a manufacturing company in the Supreme Court of Syracuse, New York."[32]

Insurers wished for more certainty in the employers' liability field. Large, unpredictable losses were common in the property and casualty insurance field from fires, floods, etc. In employers' liability insurance, however, the risks insured were not the value of a piece of property, but the shifting value of life and limbs in the courts, determined in large part by struggles in the legislative and judicial arenas. Unlike fire insurance, where rate fixing trusts developed in many states, employers' liability remained a highly competitive line of insurance never put on a firm actuarial footing.[33]

Altogether, larger jury verdicts and broadening liability became reflected in rising insurance rates and greater uncertainty for employers. Victories in the courts and legislatures translated into economic pressure on employers through the insurance mechanism and higher overall costs from work-related accidents.

The rise in the value of the right of action was less dramatic in most states. As the financial cost of a system initially subsidizing industry on the backs of injured workers began to turn into a substantial and highly uncertain factor in production, the employers' liability system became less to industry's liking. This became even more pronounced as the day to day workings of the employers' liability system exacerbated class conflict and provided a compelling moral argument against the way corporations operated. The financial cost was equalled or surpassed by a cost in class friction and the continued highly visible dramatization of murderous relations of production and subsequent cruelty in the care of injured workers and their families.

THE FAULT DISCOURSE—CLASS ANTAGONISM AND LEGITIMACY

Serious industrial accidents reveal in a grim flash the violence latent in an otherwise normalized work environment. Immediately the question arises: could the accident have been prevented, and, if so, who is at fault? An accident is a symbolic text to which different parties may give many different readings. The accident may be interpreted as the worker's fault; as an unforeseen, chance occurrence; as an intentional act by the employer in allowing predictably risky conditions to exist; or as an act of homicide. The accident demands an explanation, an account, and opens a window into broader interpretations of the conditions that gave rise to the violently inflicted bodily suffering. The accident may become the basis for various "claims": moral claims as to which parties behaved properly; claims for compensation within a market for such injuries; retributive claims demanding punishment of wrongdoers to right an injustice. Accidents become a focal point of shifting meanings attached to claims on resources for the human pain most proximately produced in the process of production.

Over time a discourse developed around attributions of fault between workers and employers for industrial accidents. This discourse was carried out by labor, business, and reformers through the courts, coroner's juries, the newspapers, and the newly created fields of accident and social research. Out of this discourse grew attributions of worker carelessness and deceit at one pole, and accusations of homicidal criminal conduct by employers at the other. In the middle were those accidents attributed to risks inherent in industry. In this view many accidents were nobody's fault and thus the extended inquiry to attribute fault was a wasteful process.

Crystal Eastman's investigation of 377 fatal work-accidents in the Pittsburgh area in 1906–1907 generated the following classification scheme: attributed solely to employer or their representatives (30 percent); attributed to those killed or fellow workmen (28 percent); attributed to both the above classes (18 percent); attributed to neither of the above classes—unavoidable accidents—(26 percent).[34] In Eastman's study employers and employees bore about equal responsibility, with chance responsible for much of the rest.

A Wisconsin study in 1907 had factory inspectors evaluate responsibility in the light of whether they thought the employer or workman had exercised "ordinary care."[35] Unlike Eastman's scheme, the Wisconsin study used the "hazard of industry" category used in an oft-quoted German study. The safety inspectors judged that over 50 percent of the industrial accidents were hazards of industry, and thus nobody's fault. The other accidents were apportioned in fault as follows: 11 percent by the employer, 24 percent by the workman, 7 percent by both, and 6 percent by fellow workers. On the other hand, a New York study concluded that only 7.6 percent of the fatal accidents studied were the responsibility of the victims, 49.5 percent the responsibility of the employer or a superior, 16 percent the shared responsibility of both, 12.2 percent the responsibility of a fellow workman or outsider, and 14.6 percent no one's responsibility.[36]

Such studies were often cited by reformers and state commissions to show that, far from being the worker's fault, work accidents were largely either the fault of the necessary hazards of industry or, secondarily, the results of negligence on the part of either the worker or the employer in roughly equal measure. Accidents just happened; both parties were often at fault, in this view, and often neither party was completely at fault or totally blameless. Basing recovery solely on employer negligence and total lack of any contributory negligence on the part of the worker would exclude large numbers of people from potential recovery or just give their claims so little bargaining leverage that they would be settled for "nuisance value" amounts.

Crystal Eastman drew the distinction between the heedless men who contributed to their fatal accidents, the inattentive ones, and the reckless:

For the heedless ones no defense is made. For the inattentive we maintain that human powers of attention, universally limited, are in their case further limited by the conditions under which the work is done—long hours, heat, noise, intense speed. For the reckless ones we maintain that natural inclination is in their case encouraged and inevitably increased by an occupation involving constant risk; recklessness is part of the trade. Not all accidents due to inattention and recklessness can be thus defended, but speaking generally, these two kinds of carelessness cannot be called faults of the workman.[37]

Risk taking, in her view, was an adaptation to the pressure and speed of work embodied in elements such as the piecework system. Similarly, Eastman felt much of the employer negligence from defective equipment was a result of the "pressure and speed of work."[38]

Many routine risks could be equated with murder. Construction accidents involving scaffolds and loose floors claimed many lives. In the view of Arthur Reeve, "In nine cases out of ten they are preventable, and are therefore little less than murder."[39] The industrial carnage of the period was such as to be easily expressed in warfare imagery; it was easy to picture the industrial labor force as a heavily risk-taking army being pushed by its officers-employers into a fearful slaughter.[40] The level of violence from instrumentalities such as coal mine cave-ins and explosions, blast furnaces, and railroads was so extreme and clearly preventable that this industrial warfare could be imagined as some sort of predictable mass murder. William Hard reported that scores of people he talked with in Chicago held the erroneous belief that the United States Steel Company plant in South Chicago secretly buried victims of accidents in its plant in mounds of slag.[41]

Coroner's jury investigations provided information for muckraking journalists of the Progressive Era who portrayed the stark contrasts of the workplace to a wide audience. William Hard's article "Making Steel and Killing Men" in *Everybody's* was perhaps the most influential journalistic piece capturing the public's attention with its juxtaposition of detail and powerful imagery. Hard evoked the gargantuan power of the steel plants, preventable violent death, and personal tragedies in a compelling package:

So came to an end the case of Ora Allen, burned to death by the slag from a pot that was being hoisted by his brother. Was it a necessary tragedy? Was all that agony, all the horror that filled the soul of Ora Allen's brother when he turned him over and recognized him, was all that wait of three days for death in the hospital, a necessary incident in the production of steel? The coroner's jury evidently did not think so, although such a jury is notably reluctant to utter a censure.[42]

Numerous commentators noted the class antagonism bred through negotiations over accidental injuries in the employers' liability system. The New York State Employers' Liability Commission noted that "from many points of view the most deplorable result of the employers' liability law is that it breeds antagonism between workman and employer."[43] Along with the antagonism generated by numerous serious accidents resulting from unsafe working conditions came a second antagonism between employers and insurers on one hand and workers and their families on the other, caused by the very structure of the employers' liability system itself.

The pain and suffering of families of injured workers became the battleground of a contest of wits and endurance between insurance companies and injured workers, each with their own sets of allies. Another element of the class antagonism bred in the process was the small amount of money received by injured workers after the long delays and coerced settlements. The lottery atmosphere was often noted: a few workers with the right injury in the right legal situation might collect large jury awards, while many with comparable injuries might receive little or nothing because of their own negligence, the involvement of a fellow servant, the legal circumstances in the jurisdiction of the case, or a hundred and one other such circumstances.

In Michigan in 1910, 7,116 fatally and nonfatally injured workers for 466 industrial employers (not including mines or railroads) received an average of $10.91 in compensation and an additional $4.39 in medical benefits.[44] Fatal accidents in these firms averaged a total recovery of $388.53 in compensation and relief; for miners the average was $1158.37. A Wisconsin study of nonfatal injuries found that about one-fourth of the workers received nothing from the employer; another two-fifths received only all or part of the medical bills, and thirty percent received something in addition to the medical bills.[45] Approximately 80 percent of the closed cases for fatal accidents in a New York study received less than $500 in compensation; over one-third of the families received nothing.[46]

Even those with large settlements were unlikely to make more than a year or two's living expenses as a net reward. Dr. R. C. Chapin estimated that an income of $800 a year in New York City was needed to maintain a normal standard of living.[47] A 1907–1908 New York study found that the average yearly wage for those fatally injured was $813.28, while the average recovery in all the settled cases was only $551.35.[48]

A study in New York on recovery of wages and medical expenses for temporary disability from industrial accidents found an average recovery by workers of 29.2 percent of their losses from employers; approximately 30 percent of this recovery came from settled lawsuits. Some 44 percent of these workers received nothing, while those with some recovery received close to half their losses. Fatal cases among married men in the New York study received an average of 17.1 percent of the sum of their medical and funeral expenses and what would have been three years' wages; 43.3 percent of these received nothing.[49] A Minnesota study of recovery for fatal accidents found that net recovery over medical, legal, and funeral expenses averaged $536, short by $1,445 of what the workers would have earned in three years of working.[50]

The system worked to provide small sums for some injuries with the prospect of large recoveries for some injuries leading employers to settle for small amounts somewhat routinely rather than risk large verdicts or invest large legal expenses to defend against small claims. While actuaries and representatives of the railroad brotherhoods estimated that only ten percent of injuries and deaths on the railroads could result in successful lawsuits, representatives of the New York Central Railroad estimated that 95 to 97 percent of the claims presented by employees were settled for some money, even if considerably less than the employee asked for.[51] Yet for employers' liability companies in New York between 1906 and 1908, only one payment was made for every eight industrial injuries reported to them.[52]

Successful suits were often appealed, causing further delays and uncertainty. Eastman estimated that of all the fatal accidents only ten percent had a chance for a successful jury verdict; recoveries in the Pittsburgh area for a married man were commonly in the neighborhood of $5,000, with 40 percent going to the attorney. Over two-thirds of the Pennsylvania Supreme Court decisions in employers' liability cases in 1904 favored plaintiffs. "A law which offers to about 10 percent of the widows and children of workingmen killed, a small chance of getting $3,000 at the end of five years, is obviously of small practical benefit to those widows and those children."[53]

Yet many of the large Pittsburgh employers provided medical benefits, employee benefit plans, and the like, responding somewhat to the potential liability of their many accidents. Including funeral expenses and employer contributions to relief associations as compensation, only 40 percent of the dependents of the 304 fatally injured workers contributing to the support of others received more than $100 in compensation. Only 4 received more than $3,000, presumably for long lawsuits. Only 30 percent received more than $500—equal to a year's wages for the lowest paid worker.[54] The value of the right of action under the employers' liability system was not simply the amounts received in court awards or settlements, but the value of the other benefits paid in relation to accidents as well. Nevertheless, although the problem was not studied in this broader context, all studies of the time showed that by any criteria, average replacement of lost earnings through lawsuits and other sources was extremely small, whether measured against lost earning power, impoverishment of the family, or pain and suffering.

Class antagonisms were further inflamed by an aspect that particularly concerned the reformers: the creation of large numbers of widows, children, and disabled workers among the deserving poor who became charity cases after they lost their income because of an industrial accident.[55] Workers and reformers shared the perception that a basic injustice existed in the gap between the worker's role in the production of wealth on one hand and the bodily and economic suffering from workplace injuries inflicted on workers and their families on the other. The severe economic losses were a measure of the low value the society placed on the workers and their families once their economically productive phase had been passed. Unlike suffering inflicted by diseases rooted in nature, industrial accidents were produced in the process of the accumulation of capital and personal wealth; they had a social location and involved choices by parties as to assuming financial responsibility or inflicting further suffering by creating an economic injury on top of a bodily one. Injuries were largely the fault of the ways employers organized the production process. Justice and money were intertwined: small and uncertain recoveries in frequently occurring tragic situations pinpointed most vividly the human waste at the core of the process of production.[56]

Many large and some medium-sized firms chose to self-insure and/or complement their handling of liability claims with an employee benefit plan. These plans often included medical, disability, and death benefits. Coverage stopped once an employee left; the plan thus served to help tie the employee to the company. Employees were sometimes the only contributors, with the employer serving just as a guarantor of the plan; sometimes the company paid a share. Death benefits in a generous plan like the one International Harvester introduced in 1910 might run to three years' wages, with a cap of $4,000. Acceptance of benefits constituted a release; the plans were expressly designed to provide a swifter benefit than would be obtained by allowing the employee and insurance company to fight it out.

ARTICULATION WITH OTHER CONFLICTS

Big business experienced a "crisis of confidence" in the early 1900s.[57] Even as their economic power grew and concentration of wealth expanded, they were losing the moral legitimacy necessary to get their way over an increasingly aroused public and their own workers. Antitrust investigations and suits, muckraker exposés of a whole range of business practices, and disasters such as the Triangle fire and the Ludlow Massacre helped create a strong negative image of big business among large sections of the public. Following ten years of prosperity, the banking panic of 1907 set off eight years of trouble in the economy, with two severe depressions (1907–1908 and 1910–1911) and several weak recoveries.[58] Businessmen faced a period of uncertainty as the anti-business climate threatened them with reforms they felt would contribute to further losses of confidence in the financial sector and lead to further economic trouble. The most popular business organization of 1908 according to Robert Wiebe, the National Prosperity and Sunshine Association, blamed the panic on "undiscriminating denunciation and legislation against capital and corporations."[59]

An upsurge of union organization and strikes ended in 1904, to be followed in the 1910s by the upsurge that David Montgomery called "the strike decade." Much of this strike activity stemmed from workers' efforts to assert craft control over the work process and to counter moves by industry such as the imposition of scientific management to change the terms of the control of work and the way effort was extracted. In 1910 almost seventy percent of the labor force worked more than 54 hours a week. The 1910s saw a wave of strikes seeking to achieve the eight-hour day.[60] Workers struggled to preserve their traditional pace of work in the face of industry efforts to promote payment schemes that undercut restrictions in output. Accidents were one clear expression of work relations that produced extreme fatigue from the combination of long hours and rapid, machine-paced, piece-work production. The struggles for shorter hours, against piecework, and for worker control of the pace of production were all part of a struggle to protect workers' lives, vitality, and bodily integrity against the demands of capital.

The strikes of the 1910s, when the compensation crisis came to a head, were often out of the control of the traditional labor leaders. "The direct, mass-involvement challenge to managerial authority and contempt for accepted AFL practice workers exhibited in 1909–1910 were to remain the outstanding characteristics of American labor struggles, not episodically but continuously for the next dozen years."[61] Direct action was not simply an IWW tactic—it was the hallmark of labor struggle in that period. In such circumstances, the added element of class conflict produced by the employers' liability system could be seen by some employers as adding fuel to an already overheated situation.

The violence of the day to day operation of production was mirrored in the violence and threatened violence of labor conflict in the period. A compensation crisis had been produced by the workings of a compensation system that allowed the employers to get off relatively cheaply with low pay-outs to the victims of workplace accidents; struggle in the courts and legislatures had helped raise the value of the injuries and the uncertainty of future liability for capital. Compensation costs were a small concern to employers compared to such issues as working hours, pay rates, and extracting greater effort from workers. The compensation crisis was a small part of the conflict between workers and employers, yet because of its symbolic importance in highlighting the worst aspects of industry and the high level of conflict at the time, it had an impact greater than if it had occurred in more placid times, such as the 1920s.

Furthermore, a European model, workers' compensation, existed that could potentially contain the crisis and provide a program that could be acceptable to capital and labor alike. The compensation crisis brought the European innovation of social insurance into the American political process as a real possibility for the first time. The compensation crisis became the focal point for a debate as to whether the American state could be modified to provide even a minimum floor to cushion the physical and financial risks of the employment relationship. As a leading symbolic issue highlighting the worst aspects of the production system, the question of whether the compensation crisis would be left to get worse or could be solved became a test of whether the political system could act to ameliorate the most glaring injustices of the production relations fueling political discontent.

CONCLUSION

Employers' liability was not simply a set of legal rules, a mixture of common law rules and employers' liability laws. Rather, it was an evolving system of relationships where attributions of fault and negotiations around the changing values of workplace injuries were played out in the shadow of widening access to juries.

Along with the class struggle over the conditions of work and the value of labor went class struggle over the conditions under which liability would be assessed for injuries incurred in the process of production and over the value of the right of action to recover damages from employers. Part of this struggle was the debate over who was at fault for the industrial carnage of the period and who would have to pay for the suffering generated in the process of creating wealth. In this class struggle workers and employers met with the aid of intermediaries: workers allied with reformers in trying to legislatively change the rules of the game, and with the newly developed plaintiffs' bar in dealing with the courts; employers played the legislative game but left most of the direct dealing with workers and their lawyers to insurance companies, whose assumption of the employers' liabilities for a price made them the central mediating force between workers, employers, and the courts. The characteristic compensation produced by the system—a few large awards, with most others receiving small settlements or nothing at all—flew in the face of what labor and the reformers forcefully asserted was massive suffering from injuries that were largely not their fault and a compensation system which compounded that suffering.

Standing at the heart of the compensation crisis was the growing access to courts and juries allowed workers by judges, the willingness of juries to find employers at fault within a set of legal rules highly favorable to employers, and the growing willingness of judges to let the awards stick. Judges and legislatures modified the early rules—which allowed workers little chance of winning—to instead at least allow workers to present more of their cases in a setting that still took a substantial showing of employer wrongdoing for them to prevail. Juries, faced with an injured worker and extremely hazardous conditions that could easily be read as employer negligence, began to raise the price of inflicting injuries upon workers to something above the level of poverty and destitution. Faced with a large uncertain risk, the insurance companies took a large profit and raised the rates to employers to a level where the costs began to be felt; uninsured employers faced the risk of bankruptcy from a large judgment. At the center of the process were Americans facing their fellows in juries across the gap created by courts and legal rules. Faced with restrictive rules within which judgments of justice could be made, Americans, through the jury form, ruled that the price for losing their limbs and lives in the production process had to rise, ruled that those who owned and controlled the workplace were often at fault and had to pay more for the suffering they created.

The employers' liability system itself became a liability to capital. The class antagonism generated in the fault debate and beating down workers in the claims settlement process weakened the fragile legitimacy of corporate capital. The rising value of the right of action necessitated growing outlays for compensation and benefit plans for a number of corporations, and created a growing uncertainty as to future costs. Rising jury awards and insurance rates created the prospects of still greater costs, the inability to find afford-able insurance, or bankruptcies due to large awards. Not only did the system inflame a class conflict already overheated; from capital's view, the system was no longer even cheap and promised to become more expensive, an-tagonistic, and uncertain. Compensation and welfare gains won by workers under the antagonistic employers' liability system could be distributed in a more predictable fashion through a less contentious no-fault system. The compensation crisis of the first decade of the twentieth century reached the point where a solution imported from Europe, workers' compensation, presented itself as a form of social insurance that could be adapted to conditions in the United States. The solution to the compensation crisis belatedly introduced the United States into the world of state-mandated social insurance.

N O T E S

1. James Weinstein, *The Corporate Ideal in the Liberal State: 1900–1918* (Boston: Beacon Press, 1968), pp. 40–61.

2. Robert Asher, "Workmen's Compensation in the United States, 1880–1935," Ph.D. dissertation, University of Minnesota, 1971.

3. Ibid., pp. 22–23.

4. I. M. Rubinow, *Social Insurance* (New York: Henry Holt and Company, 1913), pp. 166–67; I. M. Rubinow, "Medical Benefits under Workmen's Compensation," *Journal of Political Economy* 25 (June, 1917):582–83.

5. W. Jett Lauck and Edgar Sydenstricker, *Conditions of Labor in American Industries* (New York: Funk & Wagnall, 1917), p. 196.

6. Roger Lane, *Violent Death in the City* (Cambridge: Harvard University Press, 1979), p. 40.

7. Robert A. Silverman, *Law and Urban Growth* (Princeton: Princeton University Press, 1981), p. 106.

8. Lane, *Violent Death in the City*, p. 36.

9. Daniel J. Curran, "Symbolic Solutions for Deadly Dilemmas: An Analysis of Federal Coal Mine Health and Safety Legislation," *International Journal of Health Services* 14 (Spring 1984):9.

10. Lauck and Sydenstricker, *Conditions of Labor in American Industries*, pp. 192–93.

11. Ibid., p. 197.

12. Silverman, *Law and Urban Growth,* pp. 41–42.

13. George W. Alger, "The Courts and Factory Legislation," *American Journal of Sociology* 6, no. 2 (September 1900):406.

14. Gary T. Schwartz, "Tort Law and the Economy in the Nineteenth Century: A Reinterpretation," *Yale Law Journal* 90, no. 8 (July 1981):1768.

15. Wisconsin Bureau of Labor and Industrial Statistics, *Biennial Report,* 1908, p. 85.

16. Asher, "Workmen's Compensation in the United States, 1880–1935," pp. 240–41.

17. On legislative struggles for employers' liability law see Robert Asher, "Business and Workers' Welfare in the Progressive Era: Workmen's Compensation Reform in Massachusetts, 1880–1911," *Business History Review* 43, no. 4 (Winter 1969):452–75; Robert Asher, "Failure and Fulfillment: Agitation for Employers' Liability Legislation and the Origins of Workmen's Compensation in New York State, 1876–1910," *Labor History* 24, no. 2 (Spring 1983):198–222.

18. Earl R. Beckner, *A History of Labor Legislation in Illinois* (Chicago: University of Chicago Press, 1929), p. 447.

19. Asger T. Braendgaard, "Occupational Health and Safety Legislation and Working Class Political Action: A Historical and Comparative Analysis," Ph.D. dissertation, University of North Carolina, 1974, pp. 179–83, 222–26.

20. Joseph L. Castrovinci, "Prelude to Welfare Capitalism: The Role of Business in the Enactment of Workmen's Compensation Legislation in Illinois, 1905–1912," *Social Service Review* 50, no. 1 (March 1976):87.

21. Asher, "Workmen's Compensation in the United States, 1880–1935," pp. 422–23.

22. Miriam G. Abramovitz, "Business and Health Reform: Worker's Compensation and Health Insurance in the Progressive Era," D.S.W. dissertation, Columbia University, 1981, p. 203.

23. Frederick L. Hoffman, "Industrial Accidents and Industrial Diseases," *Publications of the American Statistical Association* 11, no. 88 (December 1909):599; *The Insurance Year Book 1912–1913* (Life, Casualty and Miscellaneous) (New York: Spectator, 1912), p. A-279; Roy Lubove, "Workmen's Compensation and the Prerogatives of Voluntarism," *Labor History* 8, no. 3 (Fall 1967):261. States varied in the rise of their employers' liability insurance burden on employers depending on their insurance practices, legal environment, industrial mix, and industrial growth in this decade. In Massachusetts between 1901 and 1910 premiums received by employers' liability insurers rose 140 percent; comparable figures for New York, Minnesota, and Washington were 355 percent, 410 percent, and 885 percent respectively. Compiled from *The Insurance Year Book* (Life, Casualty and Miscellaneous), published by the Spectator Company of New York, for 1900 through 1910.

24. *Report of the Employers' Liability and Workmen's Compensation Commission of the State of Michigan* (Lansing: 1911), pp. 24 & 93; *Biennial Report,* 1908, p. 31.

25. *Biennial Report,* 1908, pp. 31 & 51.

26. W. H. Allport, "Employers' Liability Insurance," *Illinois Medical Journal* 17, no. 6 (June 1910):726.

27. *Biennial Report,* 1908, p. 32.

28. Hoffman, "Industrial Accidents and Industrial Diseases," p. 599.

29. Robert Asher, "The Origins of Workmen's Compensation in Minnesota," *Minnesota History* 44, no. 4 (Winter 1974):144.

30. James L. Croyle, "Industrial Accident Liability Policy of the Early Twentieth Century," *Journal of Legal Studies* 7, no. 2 (June 1978):293–97.

31. *Report to the Legislature of the State of New York by the Commission appointed under Chapter 518 of the Laws of 1909 to inquire into the question of employer's liability and other matters, First Report*, March 19, 1910 (hereafter cited as *First Report*), p. 94.

32. Quoted in Abramovitz, "Business and Health Reform: Worker's Compensation and Health Insurance in the Progressive Era," p. 215.

33. Ralph H. Blanchard, *Liability and Compensation Insurance* (New York: D. Appleton, 1917), pp. 208–209.

34. Crystal Eastman, *Work-Accidents and the Law* (New York: Russell Sage Foundation, 1910), p. 103.

35. *Biennial Report*, 1908, p. 5.

36. *First Report*, p. 93.

37. Eastman, *Work-Accidents and the Law*, p. 95.

38. Ibid., p. 100.

39. Arthur B. Reeve, "Our Industrial Juggernaut," *Everybody's* 16, no. 2 (February 1907):154.

40. On battlefield imagery in the early 1900s see Asher, "Workman's Compensation in the United States, 1880–1935," pp. 59–60.

41. William Hard, "Making Steel and Killing Men," *Everybody's* 17, no. 5 (November 1907):590–91.

42. William Hard, "The Law of the Killed and Wounded," *Everybody's Magazine* 19 (1908):361–71.

43. *First Report*, p. 33.

44. *Report of the Employers' Liability and Workmen's Compensation Commission of the State of Michigan*, p. 9.

45. *Biennial Report*, 1908, p. 55.

46. *First Report*, p. 97.

47. *First Report*, p. 101. Average income of workingmen's families in the United States was somewhere between $700 and $800 a year. See Lauck and Sydenstricker, *Conditions of Labor in American Industries*, pp. 244–81, 354–83.

48. *First Report*, p. 101.

49. Ibid., pp. 22–23.

50. Don Dovance Lescohier, "Industrial Accidents, Employer's Liability, and Workmen's Compensation in Minnesota," Boston, American Statistical Association Publications 12 (June 1911):655.

51. Carl Hookstadt, "Comparison of Experience Under Workmen's Compensation and Employers' Liability Systems," *Monthly Labor Review* 8, no. 3 (March 1919):246.

52. *First Report*, p. 25.

53. Crystal Eastman, "A Year's Work Accidents and Their Cost," *Charities and the Commons* 21 (March 9, 1909):1170.

54. Ibid., pp. 1161–62.

55. Widows and destitute children being pushed into poverty and dependency upon charity were a common fixture of reformer and employer liability commission presentations of the industrial accident problem. In a study of families of injured workers receiving charity in New York City in 1906–1907, "trained charity agents" noted marked deterioration in about 30 percent of the families: "Chronic dependency; intemperance not before present; lowering of standards of living; breaking down in health of widow; family broken up; habit of begging developed; savings used up; furniture pawned; first experience of being dispossessed." Francis H. McLean, "Industrial Accidents and Dependency in New York State," *Charities and the Commons* 19 (December 7, 1907):1206.

56. Eastman, "A Year's Work Accidents and Their Cost," p. 1169.

57. H. M. Gittelman, "Management's Crisis of Confidence and the Origin of the National Industrial Conference Board, 1914–1916," *Business History Review* 58 (Summer 1984):153–77.

58. H. U. Faulkner, *The Decline of Laissez Faire 1897–1917* (New York: Harper & Row, 1968, originally published 1951), pp. 30–31.

59. Robert H. Wiebe, *Businessmen and Reform* (Chicago: Quadrangle, 1968, originally published 1963), p. 72.

60. David Montgomery, *Workers' Control in America* (Cambridge: Cambridge University Press, 1981), p. 96.

61. Ibid., p. 94.

CHAPTER 4   *David Rosner and Gerald Markowitz*

Safety and Health as a Class Issue: The Workers' Health Bureau of America during the 1920s

The 1920s are generally regarded as one of the most repressive eras in American labor history. As the decade opened, Eugene Debs of the Socialist Party was in prison and United States Attorney General A. Mitchell Palmer carried out his infamous Raids, resulting in the arrest and deportation of thousands of immigrant workers and political activists. As late as 1925, the New York City Police Department held half its force in reserve on May 1st to guard against a May Day uprising by "reds," and when no revolt occurred the police contented themselves with breaking up ten meetings throughout the city. For many labor organizations, the decade was nothing short of catastrophic. The IWW was essentially destroyed as an active workers' movement and, in the wake of an aggressive "open shop" campaign, the American Federation of Labor saw its membership decline from a high of 4 million in 1920 to 2.1 million in 1933. As a result of World War I the United States was converted from a debtor to a creditor nation and competed vigorously for international trade and foreign investment. The severe repression that followed the war must be understood within this framework. It was clear to management and government that, in an era of intense international competition for markets and resources, labor and the political Left's ability to disrupt "industrial progress" had to be severely curtailed.

Within this context of repression of overt political and labor activity, important organizing efforts were undertaken that had a significant impact on the lives of American workers in subsequent years. Among the most widely recognized were the rise of the American Communist Party and the efforts to organize the growing ranks of industrial workers. In addition to the successes of newer unions such as the ILGWU and the Amalgamated Clothing Workers, major organizing drives were attempted among workers in steel and automobile production.

The Workers' Health Bureau of America (1921–1928) was one important but almost totally forgotten effort by the American left to address the problems of the industrial worker.[1] Organized by a small group of radical women who sought to improve the working conditions of American workers while simultaneously aiding union organizing activities, it attracted the support of some of the most eminent public health experts and nearly 180 trade union locals throughout the country. In its eight-year history the Bureau did a series of investigations, reports, and organizing drives for a

53

wide variety of labor organizations, and it was instrumental in bringing safety and health issues to the consciousness of hundreds of thousands of American workers. In the process of organizing around safety and health, the Bureau would identify many of the political and social issues that would be subjects of debate for decades to come.

One of the long-range contributions of this organization was the development of a new conception of occupational safety and health. Unlike today's experts, who see safety and health as an engineering and technical problem that needs to be addressed by trained professionals, the Bureau saw safety and health as class issues to be controlled by unions and workers themselves. The Bureau was organized by Grace Burnham (the Executive Director) and Harriet Silverman (the Educational Secretary), two women whose roots were in the consumers' and labor movements of the Progressive Era. Burnham was a social reformer who had first gained experience in health and safety through her work with the New York State Department of Labor Factory Inspection Commission in the years after the Triangle Fire in New York City. Harriet Silverman had worked with the International Ladies' Garment Workers' Union Health Clinic, inspecting factories for health and safety violations. In 1923 they were joined by Charlotte Todes, a young Radcliffe graduate working in social welfare in Boston, who became the Organizing Secretary. Charlotte Todes (Stern) recalls Burnham was the administrator and Silverman the firebrand. They wanted "to go into workplaces, investigate working conditions, study the basis of existing hazards and then give the facts to the unions." In essence, they saw themselves as "the research adjunct to the union movement for health and safety."[2]

The Workers' Health Bureau's first pamphlet, A Health Program for Organized Labor, published in October 1921, laid out its radical ideology and its political and social commitments. Grace Burnham began by noting the new industrial conditions faced by the mass of industrial workers: "the individual is caught and bound fast in the great web of machine industry as the fly is caught in the thousand-strand web of the spider." The urban environment itself had severe and deleterious effects on the health of American citizens. "From the moment the worker starts from his tenement room, he is molded by the forces of our great cities. Crowded in subways, trains and street cars, he breathes into his lungs the foul air breathed out by hundreds of his tightly packed fellow passengers." The combination of these two new social forces—industrial production and urban living—combined in Burnham's analysis, "sapping his physical, intellectual and spiritual strength."[3]

Almost everyone could agree with this analysis of the effects on health of these new industrial and urban living conditions. But the Workers' Health Bureau offered a radically new program for their amelioration. Most scientists, industrial hygienists, and other professionals turned to the laboratory, the expert and a value-neutral, "non-politicized" science for ideas and methods to correct unhealthful conditions and depended upon employers to correct them. Burnham believed that technical personnel had to be politically committed to labor itself. The science of industrial medicine needed to be "built up carefully, slowly, scientifically; with the ideals of the laboratory," and a "thorough knowledge of working conditions" should be developed "by an industrial hygienist." Burnham saw this as only the beginning, for any such work should be conducted "FROM THE STANDPOINT OF THE WORKER." The scientist needed to be an advocate and advisor to labor, "teaching workers the meaning of the information gathered, and . . . enabling them to use this information for their own good in shop and home."[4]

The Bureau, however, went beyond a liberal position that was limited to advocacy and professional advice. It continually and pointedly illustrated the class nature of industrial disease and safety issues. "Industrial disease crushes men as effectively as the attacks of employers," Silverman pointed out in an early pamphlet on the organization's goals and direction. "Most workers' diseases are caused by his work and must be stopped at the source," she demanded. She argued that workers could not depend upon the good will and expertise of scientists, technicians, or management to solve the problems of an unsafe and unhealthy work force. Rather, she believed that the workers themselves, through organized trade unions, could eradicate hazards and improve working conditions. "Health is an industrial and class problem deserving the same place in his union program as hours, wages and working conditions," she said.

While acknowledging the benefits of the traditional union approach to health and safety matters, she criticized the fact that they focused on the symptoms rather than the causes of disability and disease. "They have provided sick and death benefits and sanitoria for the treatment of disease, usually in the advanced stages." While she acknowledged that such measures were necessary "in lessening the hardships of illness or in hastening recovery," she pointed out that "they do not cope effectively with the problem that the worker faces." The most important point that had to be recognized was "the fact that most of the worker's diseases are caused by his work and must be stopped at the source."[5] It was "[o]rganized labor, through the instrument of the trade union, [that possessed] the power to carry out a health program to prevent disease."[6]

The Bureau's insistence on trade union control of safety and health resulted in part from their less than sanguine view of state and federal governmental commitment and ability to protect the work force during the "lean years" of the 1920s. In this sense it shared the view of the AFL that "labor can not depend on 'public' agencies" to defend workers. "The sustained propaganda and lobbying necessary for the passage of some weak law limiting the use of some dangerous devices or materials shows how little public bodies are interested in industrial disease." She saw Democratic and Republican politics as undermining whatever limited advances could be achieved through such political activity.

> City, State, and National Public Health Authorities can exercise protection on the job by enforcing present laws or by introducing new laws for safe-guarding machinery, tools, materials and workplaces. . . . However, with each change in political office there is a change in policy. The moment a trade requires a drastic, swift action to introduce safe substitutes for dangerous chemicals, acids, paints or machinery, the question immediately arises, can we afford to antagonize the employers?

She concluded that "invariably, the decision is against the workers." Silverman maintained that "even if public occupational disease clinics were generally established for the examination of workers regularly, they would not only be pitifully inadequate but would suffer from the defects of all public welfare institutions—interminable red-tape—indifferent, superficial treatment—indirect contact—with the workers." If labor could not depend upon government for honest and reliable information, Silverman believed that it certainly could not depend upon more traditional research organizations such as universities or industrial foundations. "Labor needs a parallel for the Rockefeller Foundation," she argued, "AN INSTITUTE OF HEALTH RE-SEARCH."[7]

It was just this research, advisory, and advocacy role that the Bureau served. It developed a broad-based set of activities aimed at identifying, eliminating, and publicizing safety and health threats in a wide variety of worksites. One of its major functions was to study the numerous chemical- and dust-related dangers faced by trade union members. At a time when there was very little research being conducted on the effect of chemical exposure to workers, the Bureau was in the forefront of investigating the deleterious effects of such pervasive hazards as benzol, carbon dioxide, and silica dust. The Bureau also did special industry studies examining the dangers that textile workers, coal miners, glaziers, and painters faced. The results of such studies were used to work with trade unions to develop programs to correct the hazards and to help unions write health and safety provisions into their contracts. The Bureau was also active on the political front, lobbying and organizing for broader and more inclusive workers' compensation laws. Through its efforts in attracting attention to inadequate health and disease provisions in almost all state compensation acts, it was able to press some states to incorporate certain occupational diseases such as silicosis and lead poisoning. The Bureau also advocated a variety of safety and sanitary codes for different trades. For the building trades Todes wrote a code that was later approved by the New York State Federation of Labor and adopted by the New York State Department of Labor.

The Bureau saw as one of its primary roles that of an educational service for workers in their member trade unions. It therefore engaged in a widespread effort to produce pamphlets on a variety of health and safety issues for its membership. Burnham, Silverman, and Todes wrote pamphlets on such subjects as the dangers of leaded gasoline, benzol, workers' compensation, spray painting, dusty trades, bakeries, and construction. In addition, the Bureau provided practical, and sometimes impractical, advice on personal hygiene to workers on topics as varied as handwashing and home treatment of constipation.[8] But unlike the personal hygiene advice of the National Safety Council and other business-oriented groups which stressed the workers' individual responsibility for avoiding occupational diseases, the Workers' Health Bureau constantly reminded laborers of the industrial origins of their problems and the need for more permanent preventive measures—recognizing, however, that since management was unwilling to improve working conditions, such advice was necessary if workers were going to be able to protect themselves.

The Bureau undertook a number of major campaigns that established the organization as a significant force in the industrial safety and health field. The first major success of the Bureau came during its first year of existence. In July 1922 the Bureau helped six locals of New York's Journeymen, Painters and Allied Crafts organize their own Health Department, supported by union funds. They opened an office with its own laboratory and X-ray machine that served union members four evenings a week and all day Saturday. Staffed by a physician, dentist, nurse, laboratory technician, and X-ray technician, each union member was entitled to a "careful physical examination with urine analysis, blood tests, mouth examination and cleansing of the teeth."[9]

Based upon an examination of 267 men in this program, the Workers' Health Bureau, with the help of Emery Hayhurst, a respected industrial hygienist, prepared a report pointing out that although all of the men were "supposedly in good health," 39 per cent had high blood pressure, 59 per cent suffered from anemia, and four men had active lead poisoning from exposure to lead-based paints. Silverman noted that "since all but 2 per cent of the men were between 25 and 45 years of age, the percentage of arteriosclerosis—hardening of the arteries—known as the disease of *old age,* is appalling."[10] The object of this effort was not only to improve the health of individual workers but to improve the health of all workers in the painters' trade by achieving a five-day, forty-hour work week. Because of the widespread publicity that attended their report, the Painters' Union was able to win one of the first contracts guaranteeing a five-day week in New York and, later, elsewhere.[11] The contract also prohibited the use of benzene and wood alcohol in the trade.

The Bureau was committed to aiding unorganized laborers as well. In Passaic, New Jersey, during a 1926 textile workers' strike, the Bureau, with the aid of Alice B. Hamilton, conducted a health survey to help workers document the unhealthful conditions existing in the plants. Funded by the

American Fund for Public Service, the survey of 404 workers demonstrated an unusually high percentage of tubercular workers and an extremely high accident rate.[12]

The Bureau's intention to develop strong links to organized labor led it to organize a national labor health conference in Cleveland in June 1927. This conference, the first of its kind in the United States, brought together over 90 union representatives from states across the country. This two-day meeting was chaired by the President of the Washington State Federation of Labor, and papers were presented by officers of the Pennsylvania Federation of Labor; the Michigan Federation of Labor; the Auto, Aircraft and Vehicle Workers; the Bricklayers; the Painters; the Electrical Workers; and the United Mine Workers, among others. Topics included the hazards of construction work, mining, occupational diseases, national and state legislation, and data gathering.

All of the Bureau's work was closely linked to its powerful class analysis of the sources and causes of workers' diseases. Its reports on the fast-growing auto industry, for instance, placed the dangers that workers faced within the context of the industry's enormous concentration of capital and political power. Grace Burnham, in a paper read before The First National Labor Health Conference in 1927, analyzed the dangers of such concentration in "holding the power of life and death over the 25 million workers employed in factories, mills, and workshops." The automobile industry, Burnham pointed out, ranked "first of all manufacturing industries in the country in the annual wholesale value of its products." She suggested that "this concentration of capital and power in the hands of the employers breeds certain policies which are in direct opposition to the efforts of labor to protect life and health." This was done through the introduction of "high-power machinery" which "is reducing workers to mere automatons and speeding them up to the point where human endurance can go no further." In addition, the demands of modern industry "have flooded the market with an endless variety of products, with the result that chemicals, often extremely poisonous, are experimented with, regardless of their effect on the workers who are forced to handle them."[13]

Burnham recognized the central role that the growing auto industry played in the American economy, and also that the health hazards it produced went far beyond the auto plants themselves. One of the most dramatic examples of this was in 1924, when five men were killed and forty others were poisoned while manufacturing tetraethyl lead, an additive to gasoline necessary to raise its octane rating. The use of this material, still the major additive in leaded gasoline today, was opposed by the Bureau because it was a threat to both the workers who manufactured it and the larger society that used it. Due to its highly toxic nature, garage mechanics, gas station attendants, and chauffeurs were commonly understood to be in danger of lead poisoning and the horrible neurological symptoms associated with this condition. But the Bureau was also worried about the long-term effects of lead emissions, a position that won widespread approval with the introduction of unleaded gasoline a half century later (see chapter 8).[14]

Another aspect of the problem illuminated by the Bureau's political perspective was the effect on health of the changing work process in the auto industry. General Motors was partly owned by the giant Dupont Chemical Company, and the Bureau feared the enormous impact of such a concentration of capital and political muscle in imposing new and more dangerous methods of production. The Bureau was particularly worried that "The DuPont Company has been instrumental in pushing the widespread use of the spray painting machine [in the auto industry], operated under air pressure which is replacing handbrush painting and is responsible for increasing every health hazard in the industry."[15] This resulted in one disaster in the Briggs Automobile Factory in Detroit, where an explosion in the spray-painting department caused twenty-one workers to be burned to death and many others to be permanently injured.[16] The Bureau also recognized that the changing work process was responsible for more long-term dangers as well. "Twenty men formerly painted 275 chassis a day [by hand]," the Bureau pointed out. But "18 men now spray 1200. The spraying is done without the use of masks," a practice that had a disastrous effect upon the work force. "Workers spraying automobiles and railroad cars were *all poisoned in less than one year* from the time they were employed."[17]

One of the Bureau's most impressive efforts to combat long-term dangers to workers' health was in the fur hat industry. For decades, hatters had been poisoned by mercury compounds used in preparing felt hats. The expression "mad as a hatter" had its origins in this occupational disease. From its inception, the Bureau helped the United Hatters of America in Danbury, Connecticut to document the hazards that they faced. The Bureau reported that "boys 20 and 21 years old are already so badly poisoned that their hands shake continually, while many of the men who have served longer at the trade cannot even feed themselves."[18] In addition to these educational and propaganda efforts, the Bureau sought to find alternative manufacturing processes. Having heard that safe methods had been developed in the Soviet Union, Grace Burnham travelled there to meet with scientists who had found a nonpoisonous substitute for the nitrate of mercury. She "secured the formula for this substitute" and "with the cooperation of the officers of the union, the Bureau arranged with its chemist to experiment on skins sent down from Danbury until a satisfactory solution for treating the fur was found."[19] Burnham explained to C.-E.A. Winslow that subsequently the Bureau's chemist went to Danbury and showed one of the factories how to use the non-mercury "carrotting." "The detailed report of our chemist showing that the cost of making up the non-poisonous material is even less than the cost of the nitrate of mercury and his assurance that the manufacturers felt that the result was just as good if not better than the mercury skins, makes us feel that we have at last reached the point where we can make a campaign for the entire removal of nitrate of mercury in the felt hatting industry."[20] Unfortunately, this effort was made at a time when the Bureau was winding down its work, and it was unable to effect the agreement it sought with the manufacturers. But, within the course of the next fifteen years, mercury was gradually eliminated as a carrotting agent in the hat industry.

One of the things that made the Workers' Health Bureau so successful was its ability to combine a radical analysis of the problems workers faced with the most advanced and sophisticated technical methods. Their focus on health and safety allowed them to adopt the language of physicians, social workers, and even management, and thus for a time to disarm their critics. Unlike direct union organizing campaigns or wage and hour demands, workers' health and safety could be couched in non-political terms. The Workers' Health Bureau also gained legitimacy by employing highly re-garded scientists and professionals as advisors. The Bureau had, among its technical consultants, three of the nation's foremost public health experts: Emery Hayhurst of the Ohio Department of Health, C.-E.A. Winslow of Yale University, and Alice Hamilton, a pioneer in industrial medicine. Although it is doubtful that any of their consultants shared the Bureau's political commitments they generally respected the practical work that it was doing. C.-E.A. Winslow, for example, a widely respected member of the public health establishment, often congratulated the Bureau on its highly professional and technically sophisticated studies of industrial conditions. He was provided a variety of samples of toxic substances gathered by the Bureau in their plant inspections which he analyzed for them.[21]

Whatever success the Bureau was able to achieve among public health professionals, its primary constituency was the trade union movement, where it was able to attract a broad membership. It established a dues structure whereby local unions contributed $.25 per member per year, international unions and state federations paid a flat $50 per year and other District organizations paid $25 or $50 according to the size of their member-ship. Although strongest in the East, with members in New York, New Jersey, New England, and Pennsylvania, it also had support in the Midwest, the West Coast, and even Florida, Idaho, and Texas. Its Trade Union Council included the President of the Pennsylvania Federation of Labor, the Secretary of the Rhode Island Federation of Labor, and representatives of the Bakery Workers, Painters, United Mine Workers, and International Association of Machinists. Affiliated trades included auto workers, garment workers, some building trades, hatters, textile workers and others. By 1927 the Bureau had attracted the support of 190 labor organizations in 24 states, a remarkable achievement.[22] Its widespread appeal to many craft locals within the conservative American Federation of Labor is a testament to the quality of work it performed for its membership. The *Amalgamated Journal* of the iron and molders' union remarked on its influence within organized labor itself: "While the important work of the organization is to give expert advice to unions so they may incorporate in their trade agreements provisions that will guard the lives of their members and stimulate protective legislation, it is the scientific service the bureau was able to give at the very beginning that made its suggestions to union leaders carry weight."[23]

Despite the Bureau's support among American Federation of Labor locals, it quickly became apparent that its continued growth depended upon the support of William Green and the national office of the AFL. But because of its independent relationships with local unions and its activist approach, the AFL leadership was wary of becoming too closely associated with the Bureau. In 1925 delegates from the International Association of Machinists, the Stone Cutters, the Molders' Union, the Metal Polishers' Union, the American Flint Glass Workers' Union, and others, all of whom were supporters of the Bureau, introduced a resolution to the AFL convention calling for the endorsement of the Workers' Health Bureau program for health protection for trade unions. The AFL Executive Committee was opposed to the resolution and decided to refer it to the Executive Council, which referred it to the Education Committee. After the close of the convention, Green engaged in an extensive canvas of public health officials and union representatives and, despite nearly universal approval of the Workers' Health Bureau's activities, he decided to allow the supporting resolution to die in committee.[24]

Burnham, however, continued to solicit the support of the Federation, seeking an interview with Green in 1927. Unable to see him, Burnham and Silverman saw John P. Frey, a member of the Committee on Education, "with a view to working out a basis for cooperation and support." Frey agreed that the Bureau was performing a valuable service to labor and agreed to bring the issue to President Green and the Education Committee again with a view toward arriving at a definite policy of support—both financial and political—for the Bureau.

But all this activity was for naught, and the leaders of the Workers' Health Bureau began to realize that they were facing not only indifference but active hostility from the leadership of the American Federation of Labor. Beginning in 1925, Green and the AFL central office began to pressure state federations and locals to withdraw support from the Bureau. Between 1925 and 1928 the Bureau found its membership dues from locals throughout the country shrinking rapidly. In 1925–26, for example, union dues provided $6,677 for the organization. But by 1928 only $3,254 was contributed by union locals, a drop of over 50 percent. By 1928, the staff of the Bureau was very disheartened. "Two months have gone by and nothing . . . has been heard from the American Federation of Labor," the Bureau's staff complained. "The Bureau's finances are in such shape that we cannot continue operations without going heavily into debt." The staff, therefore, began to accept the idea that the work of the Bureau would have to be discontinued because of the lack of support from organized labor. "We do not feel justified in continuing to ask our affiliated unions, who represent only a small percentage of the American labor movement to assess themselves for a piece of work which can be useful to the entire membership of the American Federation of Labor and which could be easily financed with their assistance. The Administrative Staff, therefore proposes that the Bureau be discontinued."[25]

The very success that the Workers' Health Bureau was able to achieve, both programmatically and organizationally, led to its vulnerability and ultimate demise. It did not make a distinction between craft and industrial unions, helping both and even unorganized workers wherever possible. The question arises as to why the AFL hierarchy did not accept the technical and scientific research that the Workers' Health Bureau offered. It may be that the AFL feared that health and safety issues would divert attention from wages and hours in the face of open shop attacks; it may be that the AFL saw such an independent group as a threat and wanted to have exclusive control over the labor movement; it may be that the AFL was unwilling to accept the leadership and authority of an organization staffed by women; it may be that the AFL feared that the Workers' Health Bureau provided an alternative method of organizing that led workers to take a broader, industrially-based view of the work process, undermining the AFL craft-based approach.

The Workers' Health Bureau was extremely pragmatic in its approach, and this practicality led it to voluntarily disband in 1928. At the end, it was also clear that the Bureau's leadership had undergone a profound shift in their view of the relationship between government and labor. As it became obvious that organized labor was neither willing nor able to support them, the Bureau began to press government to legislate on behalf of the work force. It pushed for the expansion of the scope of workers' compensation to include blanket occupational disease laws, greater state regulations over the workplace, and safety codes. "Shall the United States stand as the only country with no National Laws for the protection of Labor?" Grace Burnham asked the large gathering of union representatives at the First National Labor Health Conference in 1927.[26] It is not that they abandoned their earlier commitment to the responsibility of trade unions. "Organized labor need not wait for legislators to act. It has the power to demand safety on the job at once. Union committees in every trade must get together to work, formulating Safety Standards, to be included in every agreement. . . . Side by side with the demand for adequate wages and shorter hours of work, must be placed the demand for safe working conditions. Every worker must . . . turn to his union for the control of these dangers."[27] Like many radicals in the labor movement of the coming Depression decade, the Bureau had begun to demand that government defend the rights of labor.

The fact that the Workers' Health Bureau of America existed for only eight years should not be taken as an indication of failure. In fact, its very success in attracting AFL locals led to its vulnerability. In addition to its numerous accomplishments, its legacy was its lasting impact on popular thinking regarding safety and health. In its brief existence, the Bureau made significant headway in promoting the notion that health and safety was an intrinsic and essential ingredient in labor demands and that concern over health and safety should be addressed through the labor movement itself. Throughout the 1930s and 1940s, such issues as benzol and lead poisoning often came to the fore in many organizing efforts by CIO-affiliated unions. Auto workers, for instance, a focus of the Bureau's concern, made safety and health conditions in the factory a central issue in their 1937 Flint sit-down strike.

It was not until the establishment of OSHA that recognition was given by government and labor to the rights of workers to a safe and healthy workplace. It is to the credit of these three women that they pointed the way.

N O T E S

1. See also Angela Nugent, "Organizing Trade Unions to Combat Disease: The Workers' Health Bureau, 1921–1928," *Labor History* 26 (Summer, 1985), 423–446.

2. Interview with Charlotte Todes Stern, January 18, 1984, New York City.

3. Grace Burnham, *A Health Program for Organized Labor*, October, 1921.

4. Ibid.

5. Harriet Silverman, *Organizing Trade Unions to Combat Disease, A New Working Class Health Program*, n.d. (ca. 1922).

6. Harriet Silverman, "The Workers' Health Bureau," *The Survey* 50 (Aug. 15, 1923): 539–40; Harriet Silverman, "Organizing Trade Unions to Combat Disease, A New Working Class Health Program," n.d. (ca. 1922).

7. Ibid.

8. Workers' Health Bureau, *Health Leaflet #1*, "Daily Health Rules for Protection Against Lead and Other Poisons," 1923: "1. Eat a Good Breakfast"; *Health Leaflet #2*, "Constipation," 1923: "Train Your Bowels to Move Regularly."

9. Silverman, "The Workers' Health Bureau."

10. Ibid.

11. "Progress Reported by the Workers' Health Bureau," *American Journal of Public Health* 15 (June 1925):569–70; Charlotte Todes Stern interview.

12. Grace Burnham, "Proposed Safeguards for the Protection of Workers in Shop Trades," paper presented to the First National Labor Health Conference, Cleveland, Ohio, June 18–19, 1927.

13. Ibid.

14. Ibid.

15. Annual Report of the Administrative Staff to the Executive Committee of the Workers' Health Bureau of America, New York, April 21, 1928, Winslow MSS, Box 102, Folder 1838, Yale University.

16. Burnham, "Proposed Safeguards for the Protection of Workers in Shop Trades."

17. Quoted in *American Journal of Public Health* 15 (June 1925):569.

18. Annual Report of the Administrative Staff to the Executive Committee of the Workers' Health Bureau of America, p. 6.

19. Grace Burnham to C.E.A. Winslow, March 30, 1928, Winslow MSS, Box 102, Folder 1838, Yale University.

20. Ibid.

21. See Workers' Health Bureau File in C.E.A. Winslow Papers at Yale University Archives.

22. *What is the Workers' Health Bureau of America* (New York, 1927); Annual Report of the Administrative Staff to the Executive Committee of the Workers' Health Bureau.

23. Charles Laue, "Workers' Health Protection to the Fore," *The Amalgamated Journal,* May 14, 1925.

24. American Federation of Labor, *Report of Proceedings of the 45th Annual Convention,* 1925; AFL Executive Committee Minutes, October 1925, p. 21, American Federation of Labor Archives, Washington, D.C.

25. Annual Report of the Administrative Staff to the Executive Committee of the Workers' Health Bureau of America, pp. 16 and 13.

26. *Proceedings, The First National Labor Health Conference,* Cleveland, Ohio, June 18–19, 1927.

27. Ibid.

Part Two

The Growth of State and Federal Involvement in Safety and Health

When Congress passed the Occupational Safety and Health Act of 1970, it was the culmination of a century of effort to force the government to assume responsibility for protecting the American work force. During the latter third of the 19th century the primary government involvement was at the state and local level. In 1877 Massachusetts passed the first factory inspection law, and during the next two decades 22 other states and a number of municipalities in the growing industrial areas passed similar acts. Around the turn of the century, beginning in mining and extending to railroads and other industries, the federal government began to play a more active role in regulating the workplace. These three chapters trace different aspects of the growing government involvement in safety and health. Jacqueline Corn examines an early effort on the state level to regulate working conditions in the coal mining industry. She shows that the horrendous accident rates in Pennsylvania mines led to organized pressure on the part of unions and others that successfully moved the state legislature to act. Rosner and Markowitz look at the growing involvement of the United States Public Health Service (PHS) and the United States Department of Labor (DOL) in safety and health between World War I and World War II. They analyze the divergent approaches of these two central federal agencies, with the PHS defining their mission as scientific observers and the DOL adopting a more activist, pro-labor orientation under the New Deal administration of Secretary of Labor Frances Perkins. Craig Zwerling examines the response of one state industrial hygiene division to beryllium poisoning, an industrial disease associated with the growing electric light industry of the 1940s. He points to the intricate relationship between industry and government that thwarted early protection of the work force. In contrast to Rosner and Markowitz, whose analysis of the state departments of labor is based upon the view from Washington, D.C., Zwerling shows, through his close inspection of one of the few state industrial hygiene divisions within a department of labor, that Washington officials' faith in these state agencies may have been misplaced. These chapters illustrate the importance of examining the activities and ideology regarding safety and health on both the state and national level.

CHAPTER **5** *Jacqueline Corn*

Protective Legislation for Coal Miners, 1870–1900: Response to Safety and Health Hazards

In the nineteenth century an ever-increasing demand for coal led to unprecedented expansion of the American mining industry. Industrial growth and technological changes had caused a rapid shift to coal for energy production in the manufacture of steel and iron, the production of steam power, and a variety of other industries including rolling mills, steamboats, stationary engines, saltworks, and glass. The total value of coal production (anthracite and bituminous) rose from $73.5 million in 1860 to $306 million in 1900.[1] Coal had supplied less than ten percent of all fuel-based energy in 1850; by 1900 it supplied approximately 73 percent of fuel-based energy.[2]

During most of the nineteenth century, practically all work in the coal mines was manual. As mechanization took place mines went deeper and grew larger. Steam engines raised up accumulated water, hoisted coal up slopes, and hauled both men and coal.[3] Technology applied to the process of extraction of coal expanded production. At the same time it created new problems of health, safety, and welfare for miners and aggravated those problems already in existence. The hazardous nature of coal mining operations had always presented problems in that industry. Working conditions, already unsafe and unhealthy, deteriorated as the mines deepened, grew larger, and employed more men.

The subject of this chapter is the response to safety and health hazards associated with United States coal mining during the last four decades of the nineteenth century. It focuses on contemporary perceptions of the relationship between the physical environment of coal mines and its effect upon the health and safety of miners, and upon responses evoked by these challenging conditions. The kinds of action or lack of action taken resulted from a combination of prevailing social attitudes toward risk associated with coal mining and the existing state of knowledge. Miners knew that deteriorating and dangerous workplaces would take their toll. The pragmatic question was how to deal with these conditions. The miners often responded with agitation for protective legislation. Indeed coal miners represented a force (albeit a weak one), when few existed, for seeking safety and health in the workplace. Neglect of health and safety issues by others reflected the social attitude that miners were responsible for their own health and welfare. Interest in miners' health and safety usually did not extend beyond the concern of a few physicians associated with the miners, rare mine inspectors, mining engineers, and the miners and their families. The neglect of coal miners' needs can also be traced in part to their geographic and social isolation.

Technological solutions for problems of the expanding industry such as ventilation, drainage, and underground transportation evolved. These innovations made possible the increased production necessary in a burgeoning industrial society. But mining engineers did not often direct advances in mining techniques toward creating a safe and healthful workplace. The coal industry's priority was to increase production to power the expanding industrial might of the United States.

HAZARDS OF COAL MINING

During the last four decades of the nineteenth century there were numerous coal mine accidents of differing severity, caused by a variety of hazards. Underground conditions also undoubtedly caused a considerable proportion of miners' diseases. Noxious substances, dusts, fumes, extremes of temperature, and over-exertion increased the risk of ill health.

Hazards included roof falls, gases which could cause explosions or asphyxiation, dust, use of explosives, fires, and haulage accidents. Utilization of new mining techniques—for example, coal cutting machines, electricity and mechanized haulage methods—increased the perils of mining. Undramatic accidents resulting from slate and roof falls, the operation of trains or cars, use of explosives, and dust diseases caused injury and death as surely as did the major explosions and disasters.

Coal mine ventilation surpassed all other areas as the most important and relevant subject in mine safety and health because it involved the supply of fresh air to working miners and animals, as well as the removal of noxious and inflammable gases. The ventilating system provided the means to circulate currents of air throughout the workings. The greater the depth of the mine, the more necessary and difficult was ventilation. Inadequate ventilation could cause explosion, fire, and asphyxiation. In deep mines, natural forces could no longer produce the circulation necessary to supply air.[4]

Risk factors related to ventilation included concentration of gases in air, dust, and the presence of a flame or spark which could create an explosion in the mine. Men died from noxious gases and lack of oxygen as well as blasts and heat. Understanding the causes of, and accepting ways to guard against, the perils of poor and inadequate ventilation spread slowly. The perception existed that explosions, fires, asphyxiation, and other hazards were the risks associated with coal mining which could be postponed, but not prevented.

The more common gases generated in coal mines were known among miners as fire-damp or choke damp, black-damp, and white-damp.[5] According to nineteenth century mining engineers, the gases emanated from a variety of sources and made ventilation a major consideration in working of mines.[6]

Lighting underground also created hazardous conditions for miners. Because of the presence of gas in mines, open lights could cause explosions. At the beginning of the nineteenth century, candles were the most convenient method of illumination in coal mines. A constant vigilance on the state of the mine was necessary because the smallest contact with an open flame was sufficient to explode certain mixtures of gases.

Throughout the nineteenth century, American mining journals consistently printed articles about illumination in mines, many dealing with safety lamps and including discussions about new inventions in that field. Although Davy invented his lamp in 1816, mining authorities discussed and explained the virtues of the Davy lamp even in the 1870s and 1880s. Electricity did not illuminate coal mines until the twentieth century.

Dust explosions in coal mines caused numerous disasters in America and England and led to investigations of the role of coal dust in mine explosions. As the mines went deeper and employed more men, and as more and more machinery found its way into the mines and caused the accumulation of dust, the number of explosions due to dust increased. Dust also caused physical impairment to the miners.

Unfortunately, it took many years for industry to define the hazards of coal dust. Mining engineers, miners, mine officials, and mine inspectors observed the explosive properties of coal dust, and individuals in England and other countries performed experiments and concluded that coal dust was explosive without any admixture of gas. But even in the 1870s, 1880s, and 1890s the view that coal dust could cause explosions was not generally accepted. Some mine inspectors and officials disputed the fact that coal dust could cause explosions and insisted that methane must be present before an explosion could be initiated.[7]

By the last decade of the nineteenth century most knowledgeable mining engineers, inspectors, and others involved in coal mining accepted the conclusion that coal dust mixed with a small amount of fire-damp (methane) or coal dust alone caused explosions. Evidence accumulated to prove that it had been a powerful factor in many of the great mining disasters. The experts agreed that coal dust promoted and aggravated explosion, and many recommended as a preventive measure that workings be sprinkled with water and that improved ways of blasting coal be sought.[8]

In deep mines, gunpowder caused explosions and fires because of the presence of either fire-damp or coal dust or both. Nevertheless, the use of some type of explosive, although unsafe, continued and caused numerous fatalities. Explosives increased dangers by adding a quick source of ignition. Blown out shots, careless handling of explosives, and the explosives themselves became possible causes of explosions. The recorded number of fatalities due to explosives in and about Pennsylvania anthracite mines from 1870 to 1900 totaled 848.[9]

Roof falls caused numerous fatal accidents in the coal mines. More men died from falls of roof and pillar coal than from any other cause between 1870 and 1899.[10] The number of fatalities grew as the use of explosives increased. Before the use of explosives the coal face advanced more slowly and exposed smaller areas of roof at a given time. The use of explosives led to a more rapid advance and exposure of larger areas of roof. Increasing depth of the mines may also have been responsible for the large number of deaths from roof falls.

Mining journals printed articles and editorials about the dangers from roof falls. A lengthy article entitled "The Causes of Accidents in Coal Mines, and the Means of Their Prevention," printed in 1875, stressed the dangers of falls of roof and made suggestions for prevention. Miners were told to make sure that they always had a plentiful supply of timber for props to support the roof and sides of working places. Experts agreed that roof falls were the greatest source of accidents in coal mines and were difficult to prevent, but they also agreed that "care and attention" paid to timbering by miners would go a long way toward preventing roof falls.

> Many, very many of these deaths occur from causes which no amount of skill or attention will obviate, but on the other hand, more rigid supervision by those in authority, and a little more display of ordinary caution on the part of the men, would doubtless do much to reduce the list in this class of accident.[11]

Many accidents in coal mines were caused by haulage operations. Removal of coal in early mining was accomplished with baskets or barrels used to hoist coal to the surface. Later, horses or mules hauled barrows or carts. By the 1870s small mine locomotives hauled coal underground; and by the end of the century horses, mules, steam locomotives, compressed air locomotives, and even some electric haulage systems were in use.

Undoubtedly, conditions in the mines were also responsible for a significant portion of coal miners' ill health. Work with harmful substances (dust, fumes, and gases) and the presence of harmful environmental conditions (heat, moisture, cold, bad ventilation, and lack of light) could cause ill health. Excessive hours of labor caused fatigue and perhaps carelessness on the job. Health hazards were misunderstood and often unsuspected. Only a handful of physicians treated miners and knew of the existence of occupational diseases. Descriptions of conditions detrimental to miners' health found in medical journals and textbooks included: impure air, believed to lead to colds, lung disorders and rheumatism;[12] deprivation of sunlight, believed to be a predisposing cause of anemia, chronic nervous irritations, tendencies toward scrofula, tubercular phthisis and allied maladies;[13] poor lighting that caused nystagmus, an eye disease;[14] and long arduous hours of labor. The medical journals of the period indicate that only a few American physicians associated coal dust with lung disease.[15] The mechanism of dust disease was not clarified, but the relationship between dust and lung disease—and between occupation and disease—was an understood, if not acted upon, fact.

Rapidly deteriorating working conditions in the mines caused the worst trials of coal miners. Early in the nineteenth century the miners worked in shallow pits or drifts with few dangers. The shift to deep mining in the fifties and sixties created new hazards; safety provisions were primitive and work became more demanding. Old methods of ventilation became inadequate, and dust posed numerous safety and health hazards. Blasting powder produced hazards and water remained a constant problem. All of these problems, plus the new impersonality between miner and operator and the rapid increase in the number of miners and mines, combined to make mining a difficult and dangerous endeavor.

PROTECTIVE LEGISLATION

In Pennsylvania, the most productive coal mining state, coal miners sought but did not secure legislation for better ventilation and mine inspection until 1869. In 1858 the number and severity of accidents in Schuylkill County, Pennsylvania caused miners to appeal to the legislature to pass a law providing for official supervision of the mines, and a bill was introduced that never even came to a vote.[16] Again in 1866 a bill entitled "An act for the protection of miners and laborers in the collieries of Schuylkill County" was introduced into the Pennsylvania legislature. The bill provided for appointment of mine inspectors and regular mine inspections. But the state legislature did not pass this bill either.[17] In 1869 the Pennsylvania State Legislature passed the Schuylkill County Ventilation Act. It provided for two important aspects of safety, ventilation and inspection. However, the Act applied to only one county.

Then, on September 6, 1869, shortly after the Schuylkill County Ventilation Act went into effect, calamity struck at the Avondale mine in Luzern County, Pennsylvania. Sparks from wood used to light the furnace which ventilated the mine set fire to the wooden works of the mine shaft. The fire burned for hours before bursting through the hoistway. It quickly filled the shaft and prevented the rescue of a single individual. One hundred seventy-nine men died of suffocation because fire blocked their only escape route.[18] A hoisting engineer at Avondale discovered the flames shooting up the shaft to the breaker. He told a newspaperman that he was startled by flames that rushed up the shaft with great fury. The fire progressed so rapidly that the engineer was only able to blow the colliery whistle to prevent a boiler explosion. In an incredibly short time everything combustible about the entire works was in flames.[19] Most accounts of the tragedy agree that the trapped miners died from suffocation, and not from fire or the gases produced by the fire.

The Avondale disaster focused attention upon mine safety and paved the way for the first significant mine safety law in Pennsylvania and the United States. Coal miners constantly faced danger, but loss of life and limb went unnoticed by the public when it happened to an individual or even a small group of men. A tragedy on the scale of Avondale could not be ignored. Avondale illustrated in a spectacular and devastating manner the increasingly hazardous conditions in coal mines. It took this major mine disaster to direct public attention to the deteriorating conditions and unsafe aspects of coal mines.[20]

After the Avondale fire, agitation for statewide safety legislation began. The following contemporary description of public response to Avondale indicates the political impact of the fire.

> The public press of the nation throughout the whole length and breadth of the land united in demanding that provisions be made by law for the application of all approved safeguards for the proper security of the lives and health of miners, and the Governor of Pennsylvania, in his annual message to the General Assembly, earnestly called the attention of that body to the necessity of a mining code to guard against the perils of the mine.[21]

Miners responded to the Avondale disaster by holding public meetings and demanding the enactment of protective laws with provisions for enforcement. John Siney, President of the Workingman's Benevolent Association, made the following statement to coal miners:

> Men, if you must die with your boots on, die for your families, your homes, your country, but do not longer consent to die like rats in a trap for those who take no more interest in you than in the pick you dig with. Let me ask if the men who own this mine would as unhesitatingly go down in it to win bread as the poor fellows whose lives were snuffed beneath where we stand and who shall henceforth live with us only as a memory. If they did, would they not provide more than one avenue of escape? Aye, men, they surely would and what they would do for themselves they must be compelled by law to do for their workmen.[22]

The *Pottsville Miner's Journal* demanded that better egress laws be passed to prevent disasters such as Avondale.[23]

Miners held meetings in the anthracite region and framed a bill for the protection of men employed in the anthracite mines of Pennsylvania.[24] Miners lobbied in Harrisburg and urged members of the state legislature to pass a bill to protect them. Petitions streamed into the legislature from the anthracite region requesting a safety act. In response to the pressure, the state legislature finally passed a safety law and Governor Geary approved it in March 1870. The new law, which protected only anthracite miners in Pennsylvania, marked the beginning of regulatory efforts to alleviate hazardous environmental aspects of coal mining in the United States.

The Act of 1870 called for complete maps of coal mines showing the progress of the workings. The law prohibited the employment of persons underground unless a mine had two outlets for each seam of coal; operators were ordered to start at once on a second opening where none existed. For the first time, a law stated that a definite minimum of pure air must be supplied to persons employed underground. It prohibited the use of furnaces in mines with breaker or chute buildings directly over or covering the top of the shaft. In mines which generated explosive gases, it provided that the mine be divided into several districts, each ventilated by a separate current of air. A mining boss was to be hired to watch over ventilation apparatus, airways, traveling ways, pumps, timbering, and signaling arrangements. The boss had to examine the workings each morning for explosive gases and keep miners from entering a dangerous zone.

The law provided for inspectors, and required that inspectors be citizens of Pennsylvania, be at least thirty years old, possess knowledge of different systems of mining, and have been connected with anthracite mining in Pennsylvania for at least five years with experience in the working and ventilation of coal mines with noxious gases. The governor would appoint inspectors for a five-year term from a list of qualified candidates supplied by a board of examiners. Inefficient inspectors could be recalled.

Inspectors' duties included: (1) examining the mines "as often as duties will permit," (2) seeing that precautions were taken to insure the safety of workmen and seeing that the law was observed, (3) proceeding to the scene of an accident and taking the necessary precautions for the safety of the men, (4) investigating causes of accidents, (5) keeping records, and (6) submitting annual reports of all accidents and of the status of mines with regard to safety and to the results of the inspectors' labor.

The law authorized inspectors to enter mines anytime during the day or night to examine anything pertaining to mining. An inspector could apply to the courts in his district to stop the working of a mine if he had no other remedy to enforce the provisions of the law.

Other safety provisions in the Act of 1870 included employment of a hoisting engineer to handle hoisting equipment, a restriction on the number of men allowed to ride on a wagon or cage at one time, control of safety lamps by the mine owner, use of signaling devices in shafts and slopes, use of safety equipment for hoisting and lowering of men, examination of boilers and machinery, and fencing off of abandoned entrances and slopes.[25]

After the Pennsylvania legislature passed the 1870 law the state of Illinois responded with mine safety laws in 1872, and Ohio did the same in 1874. Legislation for the anthracite region of Pennsylvania led to a demand for similar legislation for the bituminous region of the Commonwealth, and several attempts were made to enact safety legislation for bituminous coal mines.

Bills introduced into the Senate and House of the Pennsylvania legislature in 1871 provided for the extension of the Anthracite Act of 1870 to mines in the bituminous region. They represented the first attempt to secure a general safety law for the bituminous region of Western Pennsylvania. Petitions for safety legislation were sent to both houses and there was a similar recommendation by the governor, but the legislators would not honor these requests. Again in 1872 miners petitioned for a safety bill. Their demands resulted in the introduction into the Senate of a mining safety law. This time the legislation was blocked by coal mine operators. Instead of passing the much needed law, the legislature appointed a commission to investigate conditions in the bituminous mines of Pennsylvania.

The friends of the proposed safety measure objected to the defeating of the bill and the appointment of the commission, claiming that it was high time that such a law was enacted in as much as explosions were frequently taking place because of poor ventilation, and that the number of these explosions was increasing. It was also pointed out that the miners were agreed on this measure, having petitioned the legislature for the passage of the same, and that in fact the bill was prepared by a committee of miners. The allegation that the miners and operators were agreed that a commission should be appointed to investigate the conditions in the mines could not be true, for the miners were already fully aware of the conditions under which they were forced to work.[26]

When the Pennsylvania Bureau of Industrial Statistics investigated bituminous mines they found the following conditions:

Ventilation is in wretched condition.

Old system of ventilation used 150 years ago still in vogue. Black damp of about to 8 to 12 percent in the air which miners have to breathe in is the cause of lung diseases . . . air so bad that the common lard oil lamp dies out in it and even turpentine will not burn freely . . .

Miners have been taken out dead from the mines, after spending a night in the bad air.

Need two outlets as the mines are dangerous without them in case of explosions.

Should have a qualified person to inspect the mines and see that the dangers are removed.

Boys 8 to 9 years of age are employed in the mines, working from 10 to 12 hours a day. Many are crippled for life, as they are too young to realize the dangers in the mine.

Men work long hours, in some places between 12 and 18 hours a day.

Drinking is the result of working long hours.

Miners are agitating for the passage of a safety law the same as for the anthracite miners . . .[27]

Bituminous miners continued to work without safety legislation while the operators successfully blocked passage of such laws. The operators continually warned the legislature that if a bill became law they would be forced out of business because they could not afford the expenditure required by such an Act.

In 1875, when another investigating commission submitted its report on mining conditions, it also reported dangers to life and health in the bituminous region and recommended enactment of legislation to protect miners. The commission found that many mines did not have a system of ventilation, lacked sufficient air current to move an anemometer, and had widespread mud and water. In some mines lamps would not burn because the air contained noxious gases and occasionally commissioners could not breathe. Some mines depended on natural ventilation or on "ill contained" furnaces. Many mines did not have a means of escape in case of danger. The commission reported frequent explosions in the Monongahela River district caused by poor ventilation, and numerous accidents from roof falls, dangerous traveling ways and insufficient timber. The commissioners pointed out the absence of maps and safeguards on machinery and hoisting apparatus. They concluded that legislation was necessary to secure a reasonable extent of health and safety in the bituminous mines.[28]

Once again, a safety bill was introduced and not passed. Even though the commission had reported on the horrendous mine conditions and had strongly recommended legislation, the legislature still did not pass protective laws. The operators continued to block needed legislation, and miners continued to agitate for safety legislation. A bill introduced in 1876 was killed in conference committee.

Finally, in 1877 the state legislature passed a bill entitled, "An act providing for the means of securing the health and safety of persons employed in the bituminous mines of Pennsylvania." It contained provisions similar to those of the Act of 1870 for the anthracite mines; for example, provisions for proper ventilation, safety openings, inspections, mine bosses, and other safety measures.

The utilization and relative commercial position of anthracite and bituminous coal changed toward the end of the nineteenth century. For example, in 1875 in Pennsylvania 23,120,730 tons of anthracite coal was mined, compared to 12,443,860 tons of bituminous coal. By 1900, bituminous coal mining had exceeded anthracite; 57,367,915 tons of anthracite was mined compared to 79,842,326 tons of bituminous.[29] The dollar value of coal production also illustrates the growth of bituminous mining in Pennsylvania. In a ten-year period, 1870–1880, the dollar value of anthracite coal mined rose from approximately $38.5 million to $42.2 million, while bituminous rose from $35 million to $58.4 million.[30] This partially explains the tardiness in enacting protective legislation for bituminous miners in Pennsylvania. With greater production came more miners, more problems, eventual labor organization, and agitation for protective legislation.

Pennsylvania, the most productive of the coal mining states, had by the year 1877 some minimal safety legislation to protect coal miners. But substantial loss of life still continued in the coal mines of Pennsylvania and in the rest of the United States. Observance of the laws often fell short of their intent. In many cases, following an explosion or accident, miners claimed that inspectors had been careless or negligent. The laws were by no means perfect; they needed revision and codification. The usual pattern of response to a disaster was a call to tighten or revise the laws. Attempts at safety and health reform in the mines were thwarted time and time again. It would require passage of almost another century before the Federal Coal Mine Health and Safety Act would become law in 1969. Predictably, a tragedy of major proportions—the Farmington explosion—would trigger the passage of that legislation.

American coal miners actively participated in the search for institutionalized means to improve both working conditions and the impact of those conditions upon their own health and safety. The miners' tradition of agitation for protective legislation dates back to the nineteenth century.

It is not surprising that miners' attempts to secure protective legislation often did not succeed, given the number of other compelling problems confronting them. Along with the persistent dangers already noted, miners faced other occupationally-related problems, including the truck system, low wages, long arduous hours of work, child labor, and the system of screening and weighing coal.[31] Until recently, historians have stressed miners' response to these grievances concerning the neglect of health and safety issues. Evidence indicates that the miners knew about problems of health and safety and responded as their working conditions deteriorated and became more dangerous. Earlier discussion in this chapter of the agitation for a safety law after the Avondale fire and for protective legislation for bituminous coal miners illustrates this point. Provisions in constitutions of miners' unions also are indicative of the miners' attitudes about and need for safety legislation.

The constitutions of local labor unions formed in coal-producing areas in the late 1860s and early 1870s often contained provisions for relief of injured members and for the care of widows and orphans of miners killed in mine accidents.[32] In Pennsylvania, Ohio, Indiana, and Maryland these unions were known as Miners' and Laborers' Benevolent Associations. As conditions deteriorated and more men went into the newly deepening mines to earn their living, miners would place greater emphasis upon the need for safe and healthful mining conditions.

The Workingmen's Benevolent Association of St. Clair, incorporated in 1868, soon represented all the miners of that county and expanded to the southern and western Pennsylvania coal fields. Controversy over an eight hour law and over wages became the main issues of a strike in 1869. The Workingmen's Benevolent Association wanted a basis system which tied miners' wages to a sliding scale based upon the selling price of coal. They stated that part of their reason for seeking an eight hour day and better wages was the dangerous nature of their work. An official notice of the Workingmen's Benevolent Association read:

Resolved, taking into consideration the great risk and danger the miner and the mine laborer has to incur in pursuing his daily occupation, we claim that we should receive pay commensurate with said risk and danger, and should not be stinted down to the lowest prices given to common laborers, whose employment is safe and free from all risk to life and limb.[33]

The Workingmen's Benevolent Association agitated for the protective legislation that was passed in response to the Avondale fire.

Pennsylvania, Ohio, Illinois, Indiana, and West Virginia, the five leading coal mining states, sent delegates to a national convention at Youngstown, Ohio on October 13, 1873, and formed the Miners' National Association. A year later the president of the union stated the following objectives of the Miners' National Association: (1) to obtain legislative enactments for the more efficient management of mines, whereby the lives and health of miners may be better preserved; (2) to shorten the hours of labor in the mines; (3) to protect all branches and members when unjustly dealt with by their employers; (4) to secure the true weight of the production of the miners' materials at the mine, thus giving them and the operators their legitimate dues; (5) to sue for compensation when it is proven to the satisfaction of the Association that the negligence of the employers has been the cause of a member's death; and (6) to provide a weekly allowance for members when out of employment, in resisting any unjust demands.[34] Of the six stated objectives, three (1, 2, and 5) concerned miners' health, safety and welfare. The short-lived Miners' National Association collapsed by 1876.

Nine years passed without a national coal miners' union, although miners still had local organizations. The problems of inadequate wages, screening, semi-monthly pay, the truck system, and others still awaited solutions.

Coal miners met in September 1885 at Indianapolis, Indiana and formed the National Federation of Miners and Mine Laborers. The preamble to the new constitution reflected the miners' feeling of helplessness in the face of dangerous mining conditions, and the impersonal nature of relationships between miner and operator. The constitution of the new organization contained eight stated objectives. The following three were related to issues of health and safety: (1) The objects of this Association will be to promote the interests of miners and mine laborers morally, socially, and financially; for the protection of their health and their lives; to spread intelligence amongst them; to remove, as far as possible, the cause of strikes, and to adopt, wherever and whenever practicable, the principles of arbitration and restriction; to urge upon all miners and mine laborers the necessity of becoming citizens, that we may secure, by the use of the ballot, the service of men friendly to the cause of labor, both in our state and national legislative bodies; and to create a fund for the support and protection of the members of this Association. (2) To obtain legislative enactments for the most efficient management of mines, whereby the lives and health of our members may be better preserved. (3) To shorten the hours of labor to eight hours per day.[35]

In 1890, after years of bitter competition between the Knights of Labor and the National Federation of Miners, both groups assembled at Columbus, Ohio and adopted a constitution creating the United Mine Workers. Of the eleven stated objectives in the new constitution, five related directly to health and safety matters. They were: "(1) To secure an earning fully compatible with the dangers of our calling and the labor performed. (2) To secure

Table 5-1 Production, employees, and fatalities, showing percentage of coal-mining industry for which complete returns are available, 1870 to 1900

			PORTION OF UNITED STATES UNDER INSPECTION SERVICES OR REPORTING ACCIDENTS							
			Total for States Reporting Accidents				*Number Killed*			
			Production		Employees					
Year	Production Short Tons	Number employed	Short Tons	% of Total	Number employed	% of Total	Total	Per 1,000 Employed	Per 1,000 Short Tons Mined	Production per Death, Short Tons
1870	33,035,580		15,664,275	47.42	35,600		211	5.93	13.47	74,238
1871	46,885,080		19,342,057	41.25	37,488		210	5.60	10.86	92,105
1872	51,453,399		24,233,166	47.10	44,745		223	4.98	9.20	108,669
1873	57,602,480		26,152,837	45.40	48,199		263	5.46	10.06	99,440
1874	52,605,920		28,086,375	53.39	67,152		260	3.87	9.26	108,025
1875	52,348,320		27,350,025	52.25	85,005		269	3.06	9.51	105,192
1876	53,280,000		26,293,245	49.35	85,474		242	2.83	9.20	108,650
1877	60,501,760		30,910,316	51.09	81,142		225	2.77	7.28	137,379
1878	57,935,600		36,809,682	63.54	89,751		235	2.62	6.38	156,637
1879	68,105,799		46,447,793	68.20	96,133		317	3.30	6.82	146,523

Year										
1880	71,481,570		53,083,570	74.26	123,736		274	2.21	5.16	193,736
1881	85,881,030		56,304,138	65.56	116,128		340	2.93	6.04	165,600
1882	103,551,189		78,326,909	75.64	162,883		418	2.75	5.72	174,837
1883	115,707,525		82,356,134	71.18	162,248		542	3.34	6.58	151,949
1884	120,155,551		87,264,981	72.63	192,369		538	2.80	6.17	162,203
1885	111,160,295		92,922,981	83.59	213,178		549	2.58	5.91	169,259
1886	113,680,427		94,538,058	83.16	219,699		494	2.25	5.23	191,373
1887	130,650,511		103,774,783	79.43	228,777		504	2.20	4.86	205,902
1888	148,659,657		129,763,086	87.29	285,517		728	2.55	5.61	178,246
1889	141,229,513	311,717	127,875,451	90.54	283,198	90.85	668	2.36	5.22	191,430
1890	157,770,963	318,204	146,192,491	92.66	291,217	91.52	733	2.52	5.01	199,441
1891	168,566,669	332,147	157,654,975	93.53	310,683	93.54	956	3.08	6.06	164,912
1892	179,329,071	341,943	165,708,630	92.40	317,140	92.75	991	3.12	5.98	167,214
1893	182,352,774	363,309	177,616,520	97.40	355,091	97.74	958	2.70	5.39	185,403
1894	170,741,526	376,204	162,139,619	94.96	358,642	95.33	958	2.67	5.91	169,248
1895	193,117,530	382,879	190,104,270	98.44	376,024	98.21	1,142	3.04	6.00	166,466
1896	191,986,357	393,342	185,122,828	96.43	380,477	96.73	1,083	2.85	5.85	170,935
1897	200,229,199	397,701	194,731,837	97.25	388,585	97.71	990	2.55	5.08	196,699
1898	219,976,267	401,221	213,734,037	97.16	391,841	97.66	1,062	2.71	4.97	201,256
1899	253,741,192	410,635	243,993,172	96.16	395,607	96.34	1,241	3.14	5.08	196,610
1900	269,684,027	448,581	260,164,397	96.47	432,448	96.40	1,489	3.44	5.12	174,724

the introduction of any and all well-defined and established appliances for the preservation of life, health, and limbs of all mine employees. (3) To reduce to the lowest possible minimum the awful catastrophes which have been sweeping our fellow craftsmen to untimely graves by the thousands; by securing legislation looking to the most perfect system of ventilation, drainage, etc. (4) To enforce existing laws; and where none exist, enact and enforce them; calling for a plentiful supply of suitable timber for supporting the roof, pillars, etc., and to have all working places rendered as free from water and impure air and poisonous gases as possible. (5) To uncompromisingly demand that eight hours shall constitute a day's work, and that not more than eight hours shall be worked in any one day by any mine worker. The very nature of our employment, shut out from the sunlight and pure air, working by the aid of artificial light (in no instance to exceed one-candle power), would in itself strongly indicate that, of all men, a coal miner has the most righteous claim to an eight hour day."[36] Safety and health had found their way into the union's objectives, as evidenced by the provisions calling for use of safety appliances, the reduction of accidents, better laws concerning timbering and ventilation, and an eight hour day.

Miners, believing in legislation, and with the help of their unions, attempted to stem the tide of deteriorating health and safety conditions in the mines. It is clear that they were well aware of the need for protective legislation to correct such problems as ventilation and to reduce accidents in the mines.

By the end of the nineteenth century, twenty-one mining states had some form of safety code, protective legislation, or provision for mine inspection.[37] This legislation, often enacted in response to a crisis and distinguished by its ineffectiveness, did little to protect coal miners' health and safety. Regulation of coal mines, when it did exist, was the task of individual states, each with its own laws and administration. Coal operators often had great influence in the state legislatures and controlled the administration of safety laws. More often than not, they were more concerned with winning the coal than with the safety of their employees. Pennsylvania had appointed mine inspectors since 1870, but the state did not have an organizing or controlling head, and its mine inspectors lacked the direction of any authority. Each inspector acted according to his own judgment, without authority to enforce existing laws.

With twenty-one states having twenty-one different codes or provisions for coal mine safety, the picture remained one of disorganization and an inability to significantly decrease accidents.

Table 5-1 indicates changes in fatality statistics for the years 1870–1900.[38] It expresses fatalities per one thousand employed and per one million short tons mined, revealing that after a decrease in the number of men killed from approximately 6.0 per one thousand employed in 1870 to 3.0 per one thousand employed in 1875, the number remained remarkably constant in the vicinity of 3.0 per one thousand employed until 1900. This suggests that hazardous conditions in American mines remained remarkably constant, as judged by a fatality statistic normalized to a base of 1,000 employees. In the years 1870–1890, the number of men employed in the mines increased from approximately 35,600 to 432,118. Total numbers of deaths in the mines were 211 in 1870 and 1,489 in 1900. The average tonnage per man increased from 440 tons per year in 1870 to over 600 in 1900.

Although minimal safety legislation existed in coal mining states, health legislation had not yet been enacted. Great strides would have to be made in order to improve the health and safety of coal miners.

N O T E S

1. Howard N. Evenson, *The First Century and a Quarter of American Coal Industry* (privately printed, Pittsburgh, PA, 1942), p. 420.

2. Nathan Rosenberg, *Technology and American Economic Growth* (New York: Harper & Row, 1972), p. 159.

3. According to the *Engineering and Mining Journal*, the first mechanical methods for undercutting coal were introduced into the mines in the 1870s, and by 1880 electricity was utilized as a source of power.

4. In American coal mines, furnaces were used throughout the nineteenth century to ventilate the mines. The changeover to fans for ventilating did not occur until the twentieth century.

5. Fire-damp, perhaps the most prevalent and dangerous of the gases found in coal mines, was composed mainly of methane, sometimes called marsh gas. A mine with fire-damp was referred to by miners as a "gassy" or "fiery" mine. After-damp was the poisonous gas formed after a mine explosion. It consisted mainly of carbon dioxide. Black-damp, often called stythe, had effects similar to after-damp. White-damp produced fainting fits, giddiness, and asphyxia.

6. Andrew Roy, *Coal, A Weekly Journal of the Coal Trade* (June 6, 1883):187.

7. *Engineering and Mining Journal* (February 26, 1876 and October 1, 1881).

8. *Engineering and Mining Journal* (May 8, 1897).

9. Albert H. Fay, *Coal Mine Fatalities in the United States, 1870–1914,* Bulletin #115, United States Bureau of Mines, Department of Interior (Washington, DC: Government Printing Office, 1916), p. 79.

10. Fay, p. 16.

11. *Engineering and Mining Journal* (February 13, 1875).

12. H. van Ziemssen, *Cyclopedia of the Practice of Medicine,* vol. 19, (New York: William Wood & Company, 1879), p. 231.

13. Ziemssen, p. 233.

14. Charles Leeds Thackrah, *The Effects of Arts, Trades and Professions and Civic States and Habits of Living on Health and Longevity* (London: Longman Press, Orme, Brown Green and Longman, 1832).

15. C. E. Fowler, "The Inhalation of Foreign Substances As A Cause of Pulmonary Disease," *Occidental Health Times,* Denver, 1898; W. B. Canfield, "Report of the Section on Practice of Medicine: I. The Relation of Dusty Occupations to Pulmonary Phthisis, etc." Transaction of Medical and Chiropractic faculty, Maryland,

1889; J. T. Carpenter, "Mining Considered With Regard to Its Effects Upon Health and Life," *Tr. M. Soc. Penn.*, Philadelphia, 1869.

16. Andrew Roy, *A History of Coal Miners in the United States* (Columbus, OH: Trauger Printing Company, 1907), p. 81.

17. Alexander Trachtenberg, *The History of Legislation for Protection of Coal Miners in Pennsylvania, 1824–1915*. (New York: International Publishers, 1942), p. 30.

18. *Engineering and Mining Journal* (September 28, 1869).

19. C. M. Keenan, *Historical Documentation of Major Coal Mine Disasters in the United States—Not Classified As Explosions of Gas or Dust 1846–1962*. Bureau of Mines Bulletin #616 (Washington, DC: U.S. Government Printing Office, 1963).

20. There is a pattern, beginning with Avondale, of legislative action to protect the safety of coal miners taken only in response to a crisis. After explosions in the Orient Number 2 Mine in Frankfort, Illinois killed 119 men in 1951, Congress passed the 1952 Federal Coal Mine Safety Act. The death of 78 miners in Farmington, West Virginia provided impetus for passage of the Federal Coal Mine Health and Safety Act of 1969.

21. Andrew Roy, *The Coal Mines* (Cleveland: Robison Savage Co., 1876), pp. 180–181.

22. Trachtenberg, p. 38.

23. *Pottsville Miners' Journal* (September 9, 1869).

24. Roy, *The Coal Mines*.

25. Trachtenberg, pp. 41–47.

26. Ibid., pp. 61–62.

27. Ibid.

28. Ibid., pp. 65–66.

29. Robert D. Ballinger, *Pennsylvania's Coal Industry* (Gettysburg: Pennsylvania Historical Association, 1954), p. 38.

30. Evenson, p. 420.

31. The truck system forced miners to deal with "Company Stores," i.e. ones owned and operated by the coal company. Miners who declined to do so were discriminated against. Prices at company stores were usually higher than at other stores. The system of weighing coal deprived miners of a good part of their earnings. Cars that transported coal often varied in capacity or were improperly labeled, and the definition of what constituted a ton of coal was left up to the operator. Miners wanted both credit for all the coal they mined and just weight. Screening coal was a method of measuring the size of coal. Miners were often not paid for the smaller pieces of coal that dropped through the screen.

32. Roy, *A History of Coal Miners in the United States*, p. 68.

33. Ibid., p. 76.

34. Chris Evans, *History of the United Mine Workers of America*. (Indianapolis, 1920), vol. 1, p. 47.

35. Ibid., p. 140.

36. Evans, vol. 2, pp. 18–19.

37. The following is a list of states with some form of coal mine safety or inspection law in the 19th century, with the date the law was passed: Pennsylvania (Anthracite), 1870; Illinois, 1872; Iowa, 1873; Ohio, 1874; Maryland, 1876; Pennsylvania (Bituminous), 1877; Indiana, 1879; West Virginia, 1879; Tennessee, 1881; Missouri, 1881; Colorado, 1883; Kansas, 1883; Washington, 1883; Kentucky, 1884; Wyoming, 1886; Montana, 1889; Alabama, 1891; New Mexico, 1891; Oklahoma, 1891; Utah, 1891; Arkansas, 1893; Michigan, 1879.

38. Fay, p. 10.

CHAPTER **6**   *David Rosner and Gerald Markowitz*

Research or Advocacy: Federal Occupational Safety and Health Policies during the New Deal

In 1970, with the passage of the landmark federal Occupational Safety and Health Act, organized labor achieved a major legislative victory. The Act mandated that two agencies be established to protect the health of the American work force: the first, the Occupational Safety and Health Administration (OSHA), was to be responsible for on-site inspection, regulation, and enforcement of laws relating to dangerous and unhealthy conditions in all major industries; the second, the National Institute of Occupational Safety and Health (NIOSH), was to be responsible for establishing standards for chemical, dust, and toxic exposures to workers. OSHA was to serve policing and regulatory functions while NIOSH was to be the scientific and technical arm of this new Federal effort to protect the work force. Significantly, OSHA was placed within the Department of Labor, while NIOSH was placed within the Department of Health, Education and Welfare (now the Department of Health and Human Services)—specifically as one of the National Institutes within the Public Health Service.

On the surface, this division of responsibility between the regulatory and scientific functions of the Department of Labor and the National Institute appears non-controversial. Yet, as we will show in this chapter, this division reflects a continuing and heated debate over the proper role of federal and state governments in the field of labor relations and public health. On the most basic level the historical argument has been whether the government should be advocates for labor or technical, scientific, and objective advisors to industry and labor alike. For the most part, the U.S. Department of Labor has played an advocacy role and the Public Health Service has defined its role as that of neutral scientific advisor. At certain periods, however, when the Department of Labor, in the face of conservative business pressures, abandoned its commitment to labor, the scientists in the Public Health Service have assumed a more progressive role.

This chapter is part of a larger effort to trace the historical roots of this 1970 decision. Here, we will look specifically at the period between the two World Wars and the development of the decision to incorporate advocacy and research functions into separate agencies of government.

Between the wars, we saw a dramatic shift in the responsibility for industrial hygiene work. At the start of our period, state departments of labor were the agencies most responsible for protecting the health of workers. By the outbreak of World War II, state health departments had assumed this role. Although the implementation of policy was carried out at the state level, the ideological debates that provided the impetus for the change occurred at the federal level. Along with this shift in responsibility came new definitions of the scope and the limits of state and federal safety and health work.

The development of separate spheres for NIOSH and OSHA reflects certain long-standing differences in ideology, political commitment, and program. The differences revolved around whether the government should intervene in the "sanctity of contract" between labor and management; whether it should advocate for special interest groups; whether it should provide nothing more than technical expertise; or whether its primary goal was to enhance social and economic efficiency. Central questions were: Should information on occupational hazards gathered by scientists and technicians be used for the benefit of either management or labor? Should research be linked to reform and practical change? At what point is it the responsibility of professionals and government agencies to act? Does progressive reform depend on the police power of the state or on voluntary cooperation? Should the government provide workers and the public with information about the dangerous conditions they face? Or should it work quietly behind the scenes to correct abuses? These issues are still not resolved; they remain matters of debate within government and among public health workers.

The origins of these questions can be traced to the history of the two agencies now responsible for the health of the American work force. The U.S. Department of Labor was "created in the interest of the welfare of all the wage earners of the United States," according to its first Secretary, William B. Wilson, who was himself a former coal miner.[1] Its concern with the problems of labor led it in the early years of the twentieth century to investigate a variety of working conditions that threatened the health and safety of the work force. Its early interests included "the field of occupational diseases," and the Bureau of Labor Statistics [a division of the Department] investigated "the hygienic conditions of cotton mills; of home work; of ventilation and general sanitary conditions of clothing shops; of diseases in the glass industry; of health of women in textile factories and laundries; of poisons in the industries; of tuberculosis among wage earners; of the hygiene of the painters' trade; of anthrax as an occupational disease, etc."[2] In the early years of this century, the concerns of the Department of Labor (especially through its Bureau of Labor Statistics and its Women's and Children's Bureaus) established that working conditions and occupational safety and health were inseparable and central to labor's sphere of interest.

The Public Health Service developed with a very different set of commitments. Rather than seeing itself as an advocate for any particular interest group, the Service conceived of its role as an independent agency of scientists, physicians, and public health professionals. They shared a common interest in controlling and identifying diseases among workers, families, and communities, but they did not see themselves as advocates for the "special interests" of any one group. By the early twentieth century the Public Health Service (formerly the Marine Hospital Service) was well-known for its management of a marine hospital system, quarantine facilities in the various port cities, and laboratories for the identification of infectious diseases. Its emphasis was on the control of malaria, typhoid, yellow fever, and other diseases associated with sanitation, immigration, and commercial development. In the early years of the century, it sought to become a leader in the scientific and technical analysis of public health problems.

Beginning in the second decade of the twentieth century, the Public Health Service developed an active interest in occupationally-related diseases. In 1912, the Public Health Service was "empowered by law to study the diseases of man" and in 1915 it was given greater authority to conduct "investigations [into] occupational diseases and the relation of occupations to disease" through the organization of its Section of Industrial Hygiene and Sanitation.[3] The Public Health Service's interest in industrial hygiene stemmed from their more general interest in the spread of infectious diseases. Many public health workers believed that any future progress in the control of tuberculosis, pneumonia, and other lung disorders, for instance, demanded that attention be paid to the work environment as well as to the home environment. For those involved in the battle to improve the nation's health, it was "impossible to delimit the field of industrial hygiene as to separate it from the hygiene of the total environment."[4]

The advent of World War I made occupational safety and health a national priority. The need to conserve manpower, especially within the war-related industries, led to legislation that gave authority to the Public Health Service to protect the health of laborers in plants with government contracts. The concern over the unknown effects of new toxic chemicals such as TNT and picric acid made the Service, with its established laboratory and technical expertise, the locus of authority.

Because the Service lacked both experience in factory inspection and the legal authority to inspect plants, the PHS sought to establish a joint effort with the U.S. Department of Labor. The Labor Department's association with state and local factory inspection units, which had the statutory right to investigate conditions of work in the individual states, made such cooperation essential. A joint commission, the Working Conditions Service, was established within the Department of Labor by executive order in July 1918, staffed solely by Public Health physicians and personnel. It quickly took on the tone of a professional laboratory rather than a regulatory agency. Shortly after the end of World War I, this joint effort was abandoned. The work of toxin evaluation and identification thereafter devolved upon the Public Health Service.[5] This involvement, however, was extremely limited and consisted primarily of "investigations, leadership, advice, coordination of effort and cooperation of effort. The intention is to supplement, not to supplant, state and local government."[6]

As early as 1920 it became apparent that there was a "distinct overlapping" in function between the Public Health Service and the Department of Labor's Bureau of Labor Statistics. Therefore, Secretary of Labor Wilson called for "the coordination of the functions of these two branches of the public service" to "prevent confusion and duplication of effort."[7]

Whatever the organizational charts said, the reality was that the federal role in the control of workplace hazards remained extremely limited until the New Deal. Most authority and activity was at the state and local levels, and there the state departments of labor were the most active and involved in health and safety matters. The state departments of public health (with the notable exception of Connecticut) had little direct interest in the problems of the workplace. Traditionally concerned with problems of sanitation, food inspection, and infectious disease control, none of the health departments had active industrial hygiene divisions.

With a few notable exceptions, such as the Department of Labor in New York State, most industrial hygiene divisions did little more than occasional and routine inspections. Given the general anti-labor atmosphere of the 1920s, it is not surprising that there was a marked decline in activities to protect the work force. Unlike the Progressive Era, when state labor departments had gained tremendous authority in the wake of the agitation for the passage of workers' compensation, in the 1920s most attention was focused on industry-led and voluntary efforts to improve safety programs in individual industries.

The relative inactivity of the Department of Labor and the Public Health Service in the 1920s was in stark contrast to what was to happen in the next decade. The advent of the New Deal and the tremendous growth in federal support for industrial hygiene brought to the fore underlying tensions between labor and health departments at both state and federal levels. During the 1930s, a struggle developed over who would control the funding and administration of industrial hygiene activities at these levels. In the following pages, we will outline the general positions of the Department of Labor and the Public Health Service regarding the appropriate methods for improving laborers' health; we will then discuss the significance of Title VI of the Social Security Act, which gave funds and authority to the Public Health Service to carry on work in industrial hygiene; finally, we will describe the efforts of the Department of Labor to regain control of industrial hygiene in the months preceding World War II.

The devastation of the Depression of the 1930s, the massive unemployment, the growth in labor militancy, and the pledge of the Roosevelt

Administration to give a "New Deal" to the American people gave a new importance to the Department of Labor. The department, previously considered a second-rate backwater, achieved greater power as landmark labor legislation was enacted into law. The break with the past was symbolized by the appointment of Frances Perkins, the first woman to become a member of the Cabinet, as Secretary. She considered the inactivity of the U.S. Department of Labor under the Republican administrations of the 1920s shameful, and determined to set a new course for the Department.[8]

One of her central initiatives was the creation of the Division of Labor Standards, which combined the traditional labor concerns of hours and wages with a new focus on health and safety conditions. This integration of occupational safety and health into a broader conception of working conditions was unique to the New Deal, and was dismantled amidst corporate opposition after World War II. Much of the reason for business hostility was that the Division saw itself not simply as an advocate for labor, but in alliance with labor to improve working conditions. Verne Zimmer, the head of this new Division, explained, "the enlistment of the efforts of organized-labor groups in the interest of worker education and for the purpose of obtaining concerted action toward the adoption of improved legislation on safety and health constitutes one of the most important elements of" his new program.

Perkins herself went even further. She maintained that "legislation is not the total answer to problems such as hours, wages, sanitation, accident prevention, workmen's compensation, and unemployment insurance." She recognized that workers had to depend upon "the negotiating power of trade unions in dealing with employers. The labor unions will get as much as they can in their agreements, and if we follow with legislation to reinforce the minimal standards, we will be making definite progress."[9]

Despite the rhetoric of the department, we must recognize that many of its activities sought to integrate labor into the Democratic Party and hence diffuse what appeared in the 1930s to be a radical challenge to the status quo. Even in the context of the generally pro-labor New Deal, the Department was unwilling to impose national health and safety standards, choosing instead a voluntary approach. Responding to a proposal from Grace Abbott that the Department sponsor Federal grants in aid to state labor departments to raise health and safety standards, Verne Zimmer wrote to the Secretary, "primarily it is a question as to how far the Federal Government should go in developing a single pattern of standards and administration. . . . It seems to me that to do any real good grants for labor standards and administration would require the exercise of police power by the Federal Government, and I think this is essentially a function of the States."[10] Although such a voluntary approach might be effective in strong labor states, in most of the country where labor was weak, this method doomed workers to unhealthy and unsafe workplaces.

The coming of the New Deal also had a profound effect on the Public

Health Service's efforts to improve public health. The federal government injected massive amounts of money into state and local departments for a wide variety of services. Funds were made available for personal health care services, sanitary engineering, tuberculosis control, laboratory research, and mental hygiene as well as other services. Industrial hygiene was one area in which public health professionals sought to expand their purview. Even in the 1920s, some professional health workers sought to make industrial hygiene a central concern of health professionals. "No state health department is as active as it should be in protecting the health of workers at work," noted Henry Field Smyth at the American Public Health Association meeting in 1925.[11] And a committee on industrial hygiene of that organization recommended "that each state . . . create a position . . . or an office, bureau or division" to collect data on the prevalence of occupational diseases.[12]

Public health workers sought to develop a greater presence in this field because they saw a lack of progress in improving the health of workers during the previous decades. In 1935, the *American Journal of Public Health (AJPH)* editorialized on this very subject: "There are important exceptions, of course, to any generalization, but a 25 year retrospect of industrial health in the United States really shows very little actual accomplishment, whether measured in terms of healthier conditions of work, less illness, lessened occupational disease, a lengthening life among industrially employed." Given the fact that the public health profession had been so successful in improving general mortality statistics and health conditions during the previous three decades, it was not surprising that they sought to apply the same model to industrial hygiene work. Public health workers, impressed by the effects of scientific research and medical knowledge in improving the health of the nation, turned to the laboratory and professionals to accomplish the same thing for industrial workers. They derided what they saw as the "unscientific" approach of earlier generations of reformers: "It seems the cart has always been before the horse in industrial hygiene in this country," noted the *AJPH*. "The deleterious effects of new methods or new substances have been first discovered in the workers themselves (as the guinea pigs) . . . before scientific investigations were undertaken."[13]

The enormous concern shown by the Public Health Service also came from their recognition that industrial health was an increasingly important part of any public health effort. The change from an agricultural to a factory system over the previous decades dictated that effective health programs had to pay attention to the industrial workplace. "Advantage has been taken of the fact that the mass of our adult population is mobilized in industry and serves as an excellent group to which to bring a public health program. . . ."[14] Health workers believed that "close association with the industrial groups [must be] utilized not only to control diseases caused by industrial conditions, but to control non-industrial health afflictions and raise the general health levels of these groups."[15]

For many health professionals the reason for the lack of progress in industrial hygiene matters was the tradition of nonscientific control, by laymen and politicians, of what should have been medical and professional matters. "It is probably not going too far astray to say that much of the lack of progress in this field has been due to political maneuvers which have cleverly diverted funds . . . to lay channels . . . in place of *trained health workers*."[16] This editorial represented an implicit attack on the traditional dominance of state labor departments in industrial hygiene. One of the most venerable activists in the public health field, Emery Hayhurst, was more pointed. He lamented in 1934 that "the laws in practically every state have placed . . . health activists [interested in industrial health] in other than health departments. In two or three states this has worked out pretty well, but the careful observer knows and sees the results in all the rest. This is the Gordian knot then to be cut in America: Industrial hygiene under other than medical and health supervision."[17] Hayhurst and others believed that industrial hygiene should be a professional activity guided by physicians, technicians, and other professional personnel.

Those advocating that health departments expand their role in industrial hygiene were aided by the passage of the Social Security Act in 1935. Title VI of the Act provided for the Public Health Service to allocate funds to local and state health departments for a variety of programs, one of which was industrial hygiene. At the time, few paid attention to the fact that this item was placed in the Social Security Act. As is well known, the major provisions of the Act were to provide income assistance to the aged and the unemployed. But in the years following the passage of the Act, this money became an important lever which the Public Health Service used to prompt local and state departments of health to increase their activities in the field of occupational safety and health. In just a few years, this money was used to establish industrial hygiene divisions in a great number of state departments of health. Before 1936, only five state departments of health had industrial hygiene units. By 1938, only three years after the passage of the Act, 24 state departments had established such units.[18]

At the same time that public health departments were expanding their purview, labor departments saw their authority in this field whither away. In the 1920s, 27 labor departments supervised industrial hygiene work in their states. But after the passage of the Social Security Act, only two continued to do so. At the very same time that the federal Department of Labor was expanding its interest and activity in the field of occupational safety and health, the local departments found their funds and power eroding.

Officials involved in the U.S. Department of Labor's industrial hygiene activities were worried about the impact of Title VI. Verne Zimmer, head of the Division of Labor Standards, first warned Secretary Perkins in November of 1935 that Title VI "is going to mean that the State labor commissioners will have practically nothing to say as to the rules and regulations" regarding industrial hygiene. "On the other hand," he continued, "without financial aid few states will set up such units in the departments of labor."[19]

At the same time that tension was growing between the Labor Department and the Public Health Service, there were counter-pressures to work out an agreement for as much joint activity as possible. At the behest of President Roosevelt, they set up "Inter-Departmental Committees" to coordinate health and welfare activities, one of which was the Technical Committee on Industrial Hygiene made up of representatives of the Division of Labor Standards and the Public Health Service.[20] During the first years, the Department of Labor and the Public Health Service made a concerted effort to coordinate and organize a joint approach to the problems of industrial hygiene and safety. Originally, "it was agreed that when these new industrial hygiene units [in state health departments] are set up," Clara Beyer reported to the Secretary of Labor in 1936, "they should concentrate on service to [state] labor departments." Beyer was especially pleased that both federal departments appeared to agree that "the time had come when we should begin to apply the knowledge now available and this should be the primary function of the new industrial hygiene units." The early hope for coordination was, for Beyer, an "outstanding achievement in the development of cooperative relations."[21]

These early hopes for a harmonious working relationship between the two federal agencies were quickly dashed by continuing inter-agency rivalries both at the federal and state levels. At the federal level, Verne Zimmer sought to "persuade the Surgeon General to make . . . grants direct to the labor department of those States in which labor departments had industrial hygiene sections."[22] At the 1936 annual meeting of labor and state departments of labor officials, called by Secretary Perkins, the committee on Industrial Health and Safety proposed a resolution urging that industrial hygiene services in the states "should be directly in the departments charged with the enforcement of laws for the protection of industrial workers"—i.e., labor departments.[23] Public Health Service officials, however, did not believe that it was possible to divert funds from health departments, and sought instead to encourage cooperation at the state level. The second problem was even greater and more fundamental: the inability of some state labor and health departments to coordinate their activities. The Executive Secretary of the American Public Health Association attested to the "militant attitude" of state health officials who opposed active cooperation on the "theory that labor departments are less well equipped to handle" problems of industrial hygiene than are health departments.[24] Officials in the U.S. Department of Labor began to complain in late 1936 that despite the agreement to coordinate their efforts at the federal level, the "militant attitude" of state health officials was preventing the implementation of the agreement. A year later, Zimmer was even more wary of the growing friction that was undermining joint efforts, and by early 1939 the Department of Labor was willing to go public about its dissatisfaction with the implementation of the agreement. Verne Zimmer wrote to New York's Senator Robert Wagner that "in three large industrial states—Illinois, Pennsylvania, and North Carolina—there has developed serious friction between the state labor departments and the state health departments over conflicting approaches to this work." This friction was provoking a growing distrust by organized labor of the state health department approach to industrial hygiene, which led the federal department of labor to reevaluate its early faith in any meaningful cooperative venture.[25]

It would be possible to dismiss these conflicts as simply battles over administrative turf. But if we examine the ensuing debates over the proper role of labor and health departments in the administration of industrial hygiene work, we can see that there were fundamental ideological and political disagreements that led to this crisis. Perhaps the best way of examining these issues is to look at the debate over Senate Bill 3461 of 1939—The Murray Bill—to place safety and health activities within the purview of State labor departments.

The Murray Bill was an initiative of the Department of Labor that sought to enable "each state to provide adequate protection to workers and their families . . . by promoting the prevention and control of industrial conditions hazardous to the health of workers." The bill provided for the allocation of funds to the states with the approval of the Secretary of Labor. It further provided that the administration of these funds had to be by "the state labor department or any other agency charged with the administration of the general labor laws of the state." In practice this meant that labor departments would regain control of industrial hygiene, since they were uniformly the agencies that administered the general labor codes.

It is not surprising that the Public Health Service objected to the bill, for it was a direct attempt to retake its newly won turf. But, the Public Health Service also deeply believed that at its most fundamental level occupational safety and health was a health issue, not a labor issue. Hence, as a health issue, responsibility for its effective management should reside in the agency primarily responsible for health matters. One local health officer represented the views of the public health profession at large when he testified in opposition to the Bill. "The point that I wish to make there is this . . . industrial hygiene is not a labor problem, it is a health problem; it is a problem which requires medical knowledge and medical technique directing the effort, assisted by engineers."[26]

One of the overall goals of the Public Health Service was to reduce the loss of working time due to illness. Therefore, its investigators tended to belittle the difference between the illnesses produced by the general environment and those related specifically to the workplace. Wayne Coy, assistant administrator of the Federal Security Agency (which housed the

Public Health Service) stressed in 1940 that occupational illnesses were a relatively small part of the overall health problems facing the work force. "Today we have ample evidence of the fact that fifteen times as much time is lost by our workers from other illness as from accidents and occupational diseases combined." He felt that it was more important to evaluate the health problems of individual workers than to address the particulars of the individual plant. He called for a program to appraise "the physical defects and ill-health existing in the working population so that the health and efficiency of the individual may be improved by the correction of any defects discovered." The policy implications were "that any program for the control of accidents and occupational diseases alone will definitely fall short of the desired objectives which are the protection and improvement of the health of our workers."[27] The representatives who called for the defeat of the bill saw occupational health in largely environmental terms. Occupational diseases could not be divorced from a broader conception of the environmental causes of illness. Public health workers were strengthened in this belief by their view that the etiology of illness was unclear. To separate occupational diseases from the broader field of health was "impractical because of the often doubtful source of illness and the resulting difficulty in the delineation of jurisdiction."[28]

The Public Health Service believed that its role was to be the midwife of change, rather than the mother herself. The Service sought to study various occupational health problems and then to make their research findings available to state and local health and labor departments as well as to management. Their goal was to educate and thereby to promote change at the local level. Building on the successful public health propaganda efforts against tuberculosis and other infectious diseases, as well as on their traditional strengths in laboratory analysis and technical expertise, the Public Health Service maintained that its role was to carry on research and technical assistance tasks. In a 1937 statement of its industrial hygiene program, the PHS defined "certain minimal activities." Of the six activities, four involved conducting surveys and securing research reports on conditions in the plant. The fifth was to provide information to other state agencies, and the last was to undertake an "educational program to acquaint industry and various groups interested as to the importance of the problem."[29] Writing in the *American Journal of Public Health,* R. R. Sayers, Senior Surgeon, and J. J. Bloomfield, Sanitary Engineer of the Service, suggested that the federal government concentrate on the "collection and dissemination of information, conducting field studies, laboratory research, and protection of the health of Federal employees." They felt that the "responsibility for safeguarding the health of industrial workers rests chiefly with State and local governments." They went on to argue that this approach had been successful in alleviating the health hazards in particular industries, most notably chemicals.[30] They analyzed the past successes as a result of cooperation rather than coercion, and they placed their future hopes on a "concerted effort on the part of industry, governmental agencies, and others for an effective program in industrial hygiene."[31]

The position of the Public Health Service reflected the realities of limited federal power in the 1930s. But it also mirrored the non-confrontational approaches developed by local health officials. As one state health official asserted, "it is obviously a fact that better work can be done in the prevention of occupational diseases if you have the wholehearted cooperation of industry." Their efforts, therefore, were directed at avoiding any confrontation with business: "if industry knows that when you make a visit to their plant that any information you secure can be used as a basis of claims against it, it is perfectly obvious that the only information that you can get is that which you dig out."[32]

The voluntary, cooperative approach adopted by state health workers emanated from a recognition on their part that they lacked the authority to enter plants uninvited. Unlike most state labor departments, which possessed statutory authority to inspect factories and issue ordinances requiring improvements in unsafe or unsanitary work environments, the public health departments could only enter plants at the invitation of management. Also, health officials could only *request*, not *demand*, improvements. Perhaps because the dangers in factories were not understood to be of an epidemic nature, public health officials believed that they were powerless to prevent abuses.

This faith in voluntary cooperation by state health departments influenced and shaped policies advocated by the U.S. Public Health Service itself, which refrained from becoming involved in management-labor disputes. It even went so far as to seek exemption from court litigation. "In order to carry on [the work of research in the Service] with the minimum amount of friction and in a spirit of cooperation it may be necessary to provide a law or regulation which specifically states that the results of any investigation made by the Health department can in no way be used in litigation, either by the employer or by the employee," said R. R. Sayers, J. J. Bloomfield and W. Scott Johnson of the Public Health Service in 1934.[33]

Under Secretary Perkins, the Department of Labor represented a substantially different point of view: industrial hygiene was a labor and not a health problem. Traditionally, factory inspection programs—the regulation of work and sanitary conditions, hours of work, and fire protection—were all under the purview of state labor departments in the late nineteenth and early twentieth centuries. Both safety and health were understood to be the products of poor working conditions and inadequate protection of the work force. As Verne Zimmer pointed out at the national convention of labor officials in 1937, "you cannot separate the supervision of health in industry from that of safety. They go together."[34]

Department of Labor officials argued that it was natural for health and safety issues to be under the supervision of state labor departments because they had the legal responsibility for the enforcement of industrial codes. Labor department advocates believed that the inability and unwillingness of the Public Health Service and local health departments to enter plants at will undermined any effective industrial hygiene program. Verne Zimmer, in Congressional testimony in 1939, expressed fear that the movement of industrial hygiene departments into state and local departments of health meant that inspections and research would be conducted "only by invitation of management" and that that would "not serve the purpose [of protecting the work force] effectively." Rather than seek the voluntary cooperation of management in the inspection of plants, he maintained that industrial hygiene divisions "must be integrated and work in close cooperation with the regular inspection agency which has the legal right of entry"—i.e., the state departments of labor.[35]

Closely tied to the right of inspection went the right and obligation of enforcement. During the 1930s the health departments had assumed the responsibility for industrial hygiene without the legal right to enforce any change. At the federal level this meant that the Public Health Service defined its role solely as a research agency. At the local and state level, this meant that health departments sought voluntary and cooperative solutions to often costly and disruptive changes in the industrial workplace. As Alice Hamilton noted, research alone rarely produced effective protection of workers. "Research is very valuable, but the research that results just in the filing of reports doesn't get very far. It should be linked up to the enforcing agency. Without that it is valuable but powerless." She remarked that in most European countries industrial hygiene was usually in the sphere of labor departments.[36]

The different approaches of the two departments to the problems of industrial hygiene were the focus of a long memorandum from Frances Perkins to the administrators in the Bureau of the Budget. Perkins acknowledged that everyone was "in complete agreement" about "the need for reducing the health exposure of industrial workers in this country." But she pointed out that there were "opposing views as to the most effective method of accomplishing this end." She contrasted the "public health approach, typified by the Connecticut" health department's activities in industrial hygiene, with the "labor department approach, illustrated by the New York" Department of Labor's Industrial Hygiene Division. While both states carried on studies to determine health problems in industrial plants, Connecticut's studies were "done on invitation of the management and any recommendations about remedial measures were definitely advisory." Connecticut had "no codes or regulations dealing with specific control requirements and the hygiene unit has no administrative or regulatory powers." Connecticut had the oldest and largest health department-run industrial hygiene division, and therefore represented the state of the art in this form of organization.

Perkins contrasted this approach with that of the New York Division of Industrial Hygiene, which not only "serves industry by making plant surveys through invitation, but also initiates appraisals of industrial health hazards in connection with the regular inspection and enforcement duties of the Department of Labor." From Perkins' point of view "the outstanding thing about the New York arrangement is the fact that the Industrial Hygiene Division not only carries on exploratory and educational work, but it is provided with a regular flow of important work by reason of its integration in the Department of Labor."[37]

For Department of Labor officials, federal *regulation* of health and safety conditions was not the goal. The focuses of their activity were educational programs, publication of reports and studies, the formulation of standards, and most significantly, "the development of [state] legislation designed to prevent conditions endangering the health of workers."[38] Frances Perkins set the tone for the Department's activist stance early in her tenure as Secretary. While she valued all the research that was done by the Public Health Service and the Department of Labor, the need at the moment was for action. "In the past few years, a considerable amount of groundwork has been done in the way of valuable research and scientific study. I am anxious," she said, "to see this translated into action. . . . The need for stimulating action in controlling and preventing harmful exposure in a wide range of industrial operations can hardly be overemphasized."[39] Verne Zimmer concurred: "we have had plenty of research and scientific discussion and too little action."[40]

Their concern with protecting the work force and reforming the workplace led Department officials to turn to a different group of professionals than did the Public Health Service. Rather than depending upon medical personnel for advice and counsel, the Department of Labor turned to engineers and planners. In support of the Murray Bill, Verne Zimmer suggested that one factor that is "lost sight of in a great deal of the discussions here in these hearings [is] that engineering is the prime factor in the control of [unsafe and unhealthy] conditions."[41] In private correspondence, Zimmer was even more pointed in his criticism of the medical orientation of the Public Health Service: "I learned that the Public Health Service is about to start a study of 'Absenteeism in Industry' due to illness," he wrote to Secretary Perkins in 1935. "This is strictly our concern and I do not like their going into those things." Zimmer claimed that "this is not a medical matter, but a purely . . . industrial matter" and that "their ideas [on the subject of how to conduct an investigation of the causes of absenteeism] are simply ridiculous and run invariably to medical examination of employees."[42] Zimmer and others believed that a major problem with the scientific and technical orientation of physicians was that their interest in appropriate research designs, long-term studies, and detailed accuracy made them unable to draw conclusions about the overall relationship between working conditions and ill health. For labor officials it was essential, if they were to use doctors as technical advisors, to find some "who will come down to earth and attempt to develop practical programs for industrial disease control."[43] In contrast to the Public Health Service, Department of Labor officials believed that a medical approach limited the scope of activity too narrowly and tended to focus attention on the medical condition of individual workers rather than on the need to improve the broader work environment.

The Department of Labor's interest in the field of occupational safety and health reflected organized labor's own special interest in the subject. For decades, labor had fought for improvements in working conditions, most specifically hours of labor. Unsafe and unhealthy workplaces had always been a prime concern, especially for those unions representing the heavy industries that developed in the late nineteenth and early twentieth centuries. Miners particularly were concerned over the health hazards posed by coal and mineral dusts, explosions, and cave-ins, and had long fought for improved conditions. Dr. Walter Polakov of the United Mine Workers of America (UMW) spoke for the entire Congress of Industrial Organizations (CIO) when he said that the CIO and the UMW "have repeatedly emphasized in their resolutions and policies the paramount importance of . . . industrial hygiene," and in its 1939 convention the CIO passed a resolution urging the Department of Labor to increase its activities in the field.[44] The experience with silicosis, the devastating pulmonary condition that affected quarrymen and other laborers who drilled and used stone, proved to labor that their interests would best be protected by the Department of Labor. Throughout the 1920s and 1930s, the Public Health Service had conducted continuous and detailed studies of the debilitating effects of silicosis on quarrymen in Vermont, Missouri, and West Virginia. These were purely medical studies and were not designed for use by labor and management to eliminate the obvious hazards. It was not until the Department of Labor called conferences in 1936 and 1940 in Washington D.C. and Joplin, Missouri, that labor itself was involved in the discussion of public policy towards the protection of the work force. "To our knowledge it was the first time labor was given a chance to contribute its experience in dust control and disease prevention," claimed one of the organizers of the second conference. Elizabeth Wade White continued by saying that the organizers "cannot emphasize enough the importance of representation and participation of workers, who are exposed to the hazards of industrial disease . . . in aiding the planning and carrying out of prevention and the general industrial hygiene."[45] She went on to point out that the fantasies of medical theory often led to unrealistic and unsuccessful programs of reform when they were divorced from the practical input of workers themselves. She cited the experiences of mine, mill, and smelter workers who were told that wet (rather than dry) drilling was "a sure prevention of dust" diseases but that the president of the union and many others who had always worked with wet drilling "have advanced silicosis today."[46]

The officials in the Department of Labor shared with trade unions the belief that management would not *voluntarily* improve working conditions. There had to be some coercive pressure brought by government or labor to force management to reform. Verne Zimmer when asked by Senator Murray about the cooperation of management in improving conditions, curtly pointed out that "in the years that I was in [factory inspection] I never had a plant management meet me with open arms when I went in to pass out some orders or regulations."[47] He thought it natural that business would prefer that the Public Health Service should control industrial hygiene because they would not have any "police power," and without this power "they can't do anything to you."[48]

One of the areas of greatest struggle was over the administration of worker's compensation, the landmark legislation that guaranteed income to workers injured on the job. Whereas accidents that occurred on the job were usually compensated, compensation for occupationally-related illness was much more problematic. Even in the most liberal states a worker had to show that the illness he contracted was directly related to conditions and processes in the plant, rather than to conditions in the home or the general environment. In making such a claim, workers often turned to information gathered by government agencies on health and safety conditions in the factory. But as we have shown earlier, the Public Health Service and state departments of health were reluctant to provide specific information on the conditions in any particular plant for fear that management, upon whom they depended for access, would in the future perceive them as aligned with the laborer. They feared that providing workers with information would tend to undercut perceptions of neutrality in management-labor disputes. The labor department, however, saw workers' compensation as one of its major commitments. Therefore, they believed that industrial hygiene units had to make their studies and findings available for use in compensation claims. This put them in direct conflict with the Public Health Service. Verne Zimmer faced the issue in his Congressional testimony. He stated that "the industrial hygiene unit must be available to the State agency charged with the duty of adjudicating workmens' compensation claims. In this field, it must be ready to go into plants or places of employment to determine the nature and degree of the hazard allegedly causing disability in a claimant for workmens' compensation benefits." Not only was the industrial hygiene unit to provide information to the claimant, he said, but industrial hygiene personnel had to be "available as witnesses when needed in the prosecution of violations of existing regulations and as witnesses in workmens' compensation claims."[49] Secretary Perkins concurred. "Experienced labor law and workmen's compensation administrators will immediately recognize the vital importance of such facilities as distinguished from a purely scientific or research contribution of an industrial hygiene unit."[50]

A useful method by which to summarize the differences between the Department of Labor's Division of Labor Standards and the Public Health Service regarding their conception of industrial hygiene and occupational safety and health is to compare their model programs for the establishment of state industrial hygiene divisions. Both agreed that it was essential to identify the variety and types of health hazards that existed in various industries. The Labor Department believed that such research and investigatory activities had to be part of a strong state enforcement program, tied to the right to inspect and issue corrective orders. The Public Health Service, on the other hand, following the lead of state health departments, did not believe that enforcement powers were essential for an effective program. Instead, the state health agencies relied on the voluntary cooperation of industry. Both Federal departments agreed that it was essential that professionals be involved in the evaluation of industrial health hazards. The Department of Labor looked primarily to engineers, for they believed that correction of health hazards was intimately linked to rebuilding, reorganizing, and reforming the workplace. The Public Health Service, however, turned to medical personnel as the central professionals because they viewed industrial hygiene as primarily a medical matter. Both recognized the significance of environmental and personal factors (such as the home environment) on the health of the individual worker. But while the Department of Labor believed that their first responsibility was to correct problems at the workplace before addressing wider social causes of ill health, the Public Health Service developed a much broader model due to their own involvement in general public health and infectious disease issues. Therefore, the Public Health Service was less focused on correcting specific deficiencies of the workplace. Both recognized their responsibility as professionals, government officials, and authorities in effecting change. The Department of Labor Division of Labor Standards saw themselves as advocates of labor and primarily responsible for securing better conditions for this constituency. They believed that they could be more effective in securing change by aligning themselves with labor than by maintaining a strict neutrality. The Public Health Service, however, sought to maintain a professional distance and remained aloof from the political battles surrounding the labor movement in the 1930s. They believed that their role was to serve as neutral, objective, scientific experts.

In the end, the Department of Labor lost control over the research and scientific aspects of industrial hygiene because a *de facto* division of responsibility was accepted by Secretary Perkins. Perkins abandoned her efforts to pass the Murray Bill after President Roosevelt made it clear that he would not support the Labor Department's attempt to wrest control from the Public Health Service. Faced by the growing demands for defense expenditures and the desire to consolidate programs in the face of the.

growing threat of war, the Budget Bureau suggested to Perkins that the Murray Bill "would not be in accord with the program of the President."[51] The growing prestige of scientific medicine and the obvious successes of public health workers in the field of infectious disease control gave tremendous legitimacy to the claims of the Public Health Service that industrial hygiene was a health and not a labor issue. In light of other more pressing concerns within the labor department during the years of the Depression, we saw a *de facto* division of responsibility. In the intervening years since World War II, a great deal has occurred in the field of occupational safety and health. Yet one of the lasting legacies of this early period is the separation of technical and advocacy functions. Through the Walsh-Healy Act, the Department of Labor continued to expand its regulatory and inspection functions on the national level. Similarly, the Public Health Service, through its various state departments of health, continued to pay attention to work conditions. Yet there would rarely be any coordination between the two agencies. Because of their lack of involvement in regulation, Public Health could not develop an adequate program of industrial hygiene. Similarly, the Department of Labor would suffer from a lack of technical expertise at a time when production processes, especially in the sophisticated petrochemical industry that blossomed in the postwar years, would increasingly limit their ability to control or regulate the workplace. Thus one agency had the expertise while the other agency had the ideology and the ability to act on it. In the coming years, as occupational safety and health becomes the focus of attention once more, legislators will once again be forced to ask whether the current distinction between NIOSH and OSHA can or should be maintained.

NOTES

1. "Annual Report of the Secretary of Labor," *Reports of the Department of Labor, 1913* (Washington, DC: GPO, 1914).

2. *Reports of the Department of Labor,* 1920, pp. 230–31.

3. J. W. Schereshewsky, "Industrial Hygiene," *Public Health Reports* 30(Oct. 1, 1915):2928; "Our Public Health Service," *AJPH* 29 (Sept. 1939):1044–1045; Surgeon W. S. Bean, "The Role of the Federal Government in Promoting Industrial Hygiene," *AJPH* 15 (July 1925):626.

4. Schereshewsky, "Industrial Hygiene," p. 2930; see also L. Dublin, "Some Problems of Life Extension," quoted in *AJPH* 14 (Nov. 1924):6: "The further control of tuberculosis would seem to be largely dependent upon the development of industrial hygiene. . . . The high rates [of tuberculosis] are concentrated in a definite number of occupations which expose their workers to inorganic dusts."

5. Bean, "Role of Federal Government," p. 627.

6. Ibid.

7. *Reports of the Department of Labor, 1920,* p. 231.

8. Frances Perkins, "Oral history" (New York: Columbia University Oral History Collection) v.4, part 2, p.297: "As the head of the Department of Labor in New York," Perkins recalls in her oral history, "I knew that we had never heard from the

U.S. Department of Labor, that we never saw any of them, never heard anything from them. We didn't even get their publications unless we asked for them. In other words, there had never been the slightest effort on the part of the federal government to bring any knowledge to bear upon the problems of labor legislation in the State of New York, or . . . Massachusetts, or . . . New Jersey." Perkins saw the need to make the Federal Department more responsive to the needs of the states. "I thought it was the function of the [U.S.] Department of Labor . . . to take the lead in giving them [the state departments] the knowledge and the will to make laws of their own which were good laws." See also: Memorandum, Clara M. Beyer to the Secretary, July 16, 1936, N.A. RG 174, Department of Labor, Office of the Secretary, Folder "Labor Standards, May–Dec. 1936": "Prior to your administration, the Department of Labor limited its activities to research," Beyer noted in a memorandum in 1936. "Its responsibility as a leader in the development of a sound labor code and administrative techniques is now accepted. . . . More and more we are being called upon to advise the State labor departments as to the application of recognized methods of enforcement."

9. U.S. Department of Labor Division of Labor Standards, "Conference of Representatives of National Organizations on Cooperation in the Improvement of Labor Standards," Washington, DC, Dec. 17, 1935, N.A. Record Group 174, Department of Labor, Office of the Secretary, Folder, "Labor Standards, 1935."

10. Memorandum, Zimmer to the Secretary, Dec. 18, 1936, N.A. R.G. 100, 0-0-11, Box 13: "I realize that our present process of education and persuasion is slow, but I think it more sound and much less dangerous than the recourse to the grants-in-aid idea."

11. Henry Field Smyth, "Recent Developments in Industrial Hygiene," *AJPH* 16 (Feb. 1926):127.

12. "Report of the Committee on Industrial Hygiene," *AJPH* 14 (Oct. 1924):854.

13. Editorial, *AJPH* 25 (July, 1935):859.

14. Federal Security Agency to Murray, May 8, 1940.

15. Dr. A. S. Gray, Director, Bureau of Occupational Diseases, 76th Cong., 3 R. D. Sess., Senate, Committee on Education and Labor, Hearings on s.3461, "Prevention of Industrial Conditions Hazardous to the Health of Employees," May 13–15, 1940 (Washington, 1940) pp. 12–13. (Hereafter referred to as The Murray Bill).

16. E.g. *AJPH* 25 (July, 1935):859.

17. Emory R. Hayhurst, "The Industrial Hygiene Section, 1914–1934," *AJPH* 24 (Oct. 1934):1041–42.

18. Thomas Parran to Verne Zimmer, Dec. 8, 1938, NA, RG 90, "General Classified Establishments;" Memorandum, Dr. Jones to Mrs. Beyer, Jan. 11, 1937, NA, RG 100, first series, 1934–1937, 7-2-3-5, Box 65; Clara M. Beyer, "Problems, Needs and Possibilities in the Field of Public Health and Safety Education," NA RG 100, second series, 1937–1941, 7-0-7-3-2, Box 34.

19. Memorandum, Zimmer to the Secretary, Nov. 23, 1935, N.A. Department of Labor, Office of the Secretary, Folder, "Division of Labor Standards, 1935."

20. Summary of work of the Technical Committees, March 1, 1937, N.A. Department of Labor, Office of the Secretary, Folder: Inter-Departmental committee to Coordinate Health and Welfare Activities; Press Release, "Statement by the President," Aug. 15, 1935, N.A., R.G. 100, 7-0-7-3-2, Box 34.

21. Memorandum, Beyer to the Secretary, April 14, 1936; "Statement by the President," Aug. 15, 1935.

22. Memorandum, Zimmer to Lubin, Sept. 30, 1936, N.A., R.G. 100, 7-0-7-4-1, Box 58.

23. Third National Conference on Labor Legislation, Nov. 9–11, 1936.

24. Atwater to Jones, Oct. 2, 1936, enclosed in Zimmer to Lubin, Sept. 30, 1936.

25. Zimmer to Frank C. MacDonald, Nov. 13, 1936; Zimmer to W. H. Young, Nov. 4, 1937; and Zimmer to Wagner, Jan. 26, 1939, N.A. R.G. 100, 0-3-0-1, Box 13.

26. Albert E. Austin, Fourth Congressional District—Conn., (The Murray Bill), p. 33.

27. Wayne Coy, Murray Bill, pp. 104–105.

28. Ibid.; see also Acting Administrator, Federal Security Agency to Senator Murray, May 8, 1940, N.A. Record Group 90, General Files: "Thus it is proposed to take from this field that part which relates to occupational disease and industrial hazards and to make special provision for it entirely divorced from the broader field to which industrial hygiene efforts are usually directed."

29. U.S. Treasury Department, Public Health Service, "Evaluation of the Industrial Hygiene Problems of a State," *Public Health Bulletin #236* (Washington, DC: GPO, 1937), pp. 56–57.

30. R. R. Sayers and J. J. Bloomfield, "Industrial Hygiene Activities in the United States," *AJPH* 26 (November, 1936):1087.

31. Federal Security Agency, U.S. Public Health Service, "A Preliminary Survey of the Industrial Hygiene Problem in the United States," *Public Health Bulletin #259* (Washington, DC: GPO, 1940), p. 66.

32. Dr. A. S. Gray, Director, Bureau of Occupational Disease, State Department of Health—Conn., Murray Bill, p. 19; one of the most revealing statements of the perceived powerlessness of the public health establishment can be found in correspondence between Dr. J. Schereschewsky, Assistant Surgeon General of the Public Health Service, and Dr. Lanza, director of their field investigations. Dr. Lanza wrote to Dr. Schereschewsky in July 1918 asking "where we stand in the matter of authority; that is to say, when we send a copy of the inspection report containing specific recommendations for improvement to the plant concerned, are we in a position to include in that report a letter asking or directing the heads of the plant to carry out the recommendations?" Lanza feared that "there will be a goodly number where the effect [of plant inspections and recommendations] will not be permanent . . . I do not think that much would be accomplished by sending copies of reports to the State Boards of Health or the state Boards of Labor. . . ." Lanza clearly was asking for the Service to assume a regulatory and policing role. Dr. Schereschewsky replied that "I may say that the Public Health Service has no specific authority to compel any immediate changes in plants which are investigated. The only course of action open to us is to make necessary recommendations and to urge that they be put into effect." He pointed out that the "only other course open to us at the present time consists in furnishing the labor department of the State concerned with a report of the investigations." Lanza to Dr. Schereschewsky, July 17, 1918, and Schereschewsky to Lanza, July 22, 1918, N.A. Record Group 90.

33. U.S. Treasury Department, Public Health Service, "The Potential Problems of Industrial Hygiene in a Typical Industrial Area in the United States," *Public Health Bulletin #216* (Washington, DC: GPO, 1934), p. 32.

34. Proceedings of the 23rd Convention of Government Labor Officials, Toronto, Canada, Sept. 1937, in "Labor Laws and Their Administration," 1937, U.S. Department of Labor, Bureau of Labor Statistics, Bulletin #653 (Washington, DC: GPO, 1938), p. 196; Secretary Perkins noted that "it is significant that in practically every state authority to inspect places of employment and the duty to enforce all existing laws and regulations in respect to safety and health in those places is invariably invested in State labor departments." Department of Labor officials were contemptuous of the health officers' lack of elementary knowledge concerning factory conditions in their districts; Perkins to Bureau of the Budget, April 25, 1940, N.A. Record Group 90, Public Health Service, General Files, p. 2.

35. U.S. Congress, Senate, 76th Congress, First Session, "Hearings Before a Sub-Committee of the Committee on Education and Labor to Establish a National

Health Program", S.1620 (Washington, DC: GPO, 1939), p. 867. Hereafter referred to as the National Health Bill.

36. Murray Bill; Labor officials also objected to the Public Health Service's position that police power was not necessary for running an industrial hygiene unit. They held that the ideology of voluntarism and cooperation between industry and government was either naive or dishonest. The head of New York State's Division of Industrial Hygiene, one of the few units still located within the State Labor Department, was extremely pointed in his criticism of what he saw as a hypocritical and specious argument of the Public Health Service. "Anybody who says that a health department does not need police power does not understand what the public health law is based upon, because it is based upon the police power of the state." He pointed out that health departments had used their powers in many other instances and only here refused to acknowledge their importance: National Health Bill, p. 258

37. Frances Perkins to Bureau of the Budget, April 25, 1940, N.A. Record Group 90, Public Health Service, General Files.

38. Secretary of Labor, *27th Annual Report*, 1939, p. 57

39. Perkins to Josephine Roche, Nov. 11, 1935, N.A. Record Group 90, Public Health Service, State Boards of Health: "Knowledge of the harmful effect on workers of certain dusts, fumes, gases, and of various solvents, solutions, and contaminants is widespread among industrial hygienists, physicians and chemical engineers." She noted that "literally tons of literature" was available on these hazards and "as a result there is today a plethora of data but a dearth of practical application of this knowledge in terms of actual engineering control." Lest anyone misunderstand her position she concluded by saying that "in other words, the major need at present is not one of additional medical research but a wider use of existing knowledge about these health hazards": Perkins to Bureau of the Budget, p. 2.

40. Memorandum, Verne Zimmer to the Secretary, Nov. 1, 1935, N.A. Record Group 174, Department of Labor, Office of the Secretary, Folder, "Labor Standards, 1935."

41. Murray Bill, p. 109.

42. Zimmer to the Secretary, Sept. 27, 1935, N.A., Record Group 174, Department of Labor, Office of the Secretary, Labor Standards, 1935.

43. Memorandum, V. A. Zimmer to the Secretary, Nov. 1, 1935. N.A. Record Group 174, Department of Labor, Office of the Secretary, Labor Standards, 1935. Matthew Woll, Chairman of the Committee on Social Security of the American Federation of Labor (AFL), agreed with the CIO in this matter. "We respectfully point out and insist that the administration of any part [of a national health program] which involved the rights of labor should be lodged in the United States Department of Labor and the various State departments. This is especially true of industry hygiene. . . ." He even felt that occupational diseases that had traditionally been in the purview of health departments "should be administered by the labor department instead of by the health department": Murray Bill, p. 65.

44. National Health Bill, pp. 229, 230.

45. Murray Bill, pp. 63–64.

46. Ibid.

47. National Health Bill, p. 868.

48. Murray Bill, p. 115.

49. National Health Bill, p. 867.

50. Perkins to Bureau of Budget, April 25, 1940.

51. Assistant director, Bureau of the Budget to Perkins, March 8, 1941, N.A. R.G. 100, 7-0-7-4-1, Box 38; see also Memorandum, Reilly to Zimmer, March 11, 1941: "I take it that the nub of the last paragraph is that the bill was invading a domain reserved for the Public Health Service."

Salem Sarcoid: The Origins of Beryllium Disease

Controversies are not unusual among scientific researchers. In some cases, the argument centers on priority of discovery among individual researchers or research institutions. Other disputes revolve around different theoretical paradigms within a scientific discipline.[1] Still other times, arguments arise when different researchers from different disciplines approach problems at their common boundaries.

Although similar controversies arise in the study of occupational disease, the role of social and economic contexts—especially the struggle between labor and capital—are particularly important. Technologic change and occupational disease arise within the context of labor conflict. Because knowledge of occupational disease has implications not only for workers' health, but for capitalist profits, the study of these diseases is often very controversial. Management and its allies generally argue that a substance is not toxic or is less toxic, while labor and its allies argue that the same substance is more dangerous.

This conflict is reflected in both regulatory agencies and academic institutions. Different professional organizations acquire reputations for being sympathetic to either labor or management, as do different academic programs. Indeed, individual researchers are often perceived as being in one camp or the other.

The history of an epidemic of respiratory disease related to the use of beryllium in the manufacturing process at a fluorescent light plant in Salem, Massachusetts, provides an opportunity to see this conflict at work within the Division of Occupational Hygiene of the Massachusetts Department of Labor. The background and social commitments of the researchers played an important part in determining the timing of the public identification of beryllium disease.

THE FLUORESCENT LAMP INDUSTRY

To understand the emergence of fluorescent lights and the hazards they posed to workers, we must first look at the electric lamp industry of the 1930s and 1940s. During the 1930s the industry was dominated by the General Electric Company, which produced almost 60 percent of the tungsten filament lamps marketed in the United States (Table 7-1). Because of the licenses on its patents, General Electric exerted even greater market control than Table 7-1 suggests. General Electric had two types of licenses,

Table 7-1 Approximate Shares of U.S. Market for
Large Tungsten-Filament Lamps, 1937

Supplier	Percent Market Share
General Electric	59.3
Westinghouse	19.0
Sylvania	4.4
Consolidated	2.8
Ken-Rad Tube	1.1
Other Domestic	8.8
Imported	4.6
TOTAL	100.0

Source: Arthur A. Bright, Jr., and W. Rupert MacLaurin, "Economic
Factors Influencing the Development and Introduction of the Fluorescent Lamp," *Journal of Political Economy* 51:431, 1943.

A and B. The only A licensee was Westinghouse. As an A licensee, Westinghouse was allowed to use the "Mazda" trademark. Its sales quota was set at
34.12 percent of General Electric's sales. The royalty rate on the quota was
only 1 percent, but above the quota it was 30 percent. Sylvania, Con-
solidated, and Ken-Rad were B licensees with sales quotas of 9.12 percent,
3.89093 percent, and 1.7584 percent of General Electric's sales respectively.
They paid 3 percent royalty on the quota and 20 percent above the quota.
Thus, General Electric through its licensees controlled 86.6 percent of the
American market. It is not surprising that General Electric did most of the
early research on fluorescent lamps—their size gave them superior research
facilities.[2]

The idea of a fluorescent lamp was not new. As early as 1859, Alexander
Edmond Becquerel had described low-pressure discharge tubes containing
luminescent solids in powder form. Over the years, much experimentation
had been done, but mostly in Europe. It was not until 1935 that General
Electric began to develop a fluorescent lamp for the American market, and
then only after their consultant, Nobel prize winner Arthur H. Compton,
had seen fluorescent lamps in Europe and extolled them as the "lighting of
the future."[3]

Fluorescent lamps are somewhat more complicated than their incandescent predecessors. An incandescent lamp is simply a tungsten filament in an evacuated bulb. As the current passes through the filament, it is
heated and gives off light. A fluorescent lamp is a long tube at low pressure,
filled with a drop or two of mercury. An electrical discharge is passed
through the mercury, which gives off ultraviolet light. The tube is lined with
a fluorescent powder which emits visible light when stimulated by the
ultraviolet discharge. Although fluorescent lamps were large and costly, they

had one major advantage over their incandescent predecessors: efficiency. A 40-watt fluorescent lamp produced 52 lumens per watt and had a life expectancy of 2,500 hours; a 40-watt frosted incandescent lamp produced 11.7 lumens per watt and had a life expectancy of 1,000 hours.[4]

Although General Electric developed the first American fluorescent lamp, they were not anxious to market it aggressively. As we have seen, General Electric already controlled the market for incandescent lamps. The new fluorescent lamp would only compete with General Electric's own product. Moreover, General Electric had a long and close working relationship with the utilities. And the utilities were concerned that the increased efficiency of fluorescent lamps would lead to a decrease in wattage. Therefore, both the utilities and General Electric favored a slow introduction of fluorescent lamps. They also wanted to be sure that the new lamps were thoroughly tested so that they did not jeopardize the reputation for high quality that incandescent lamps had developed over the years.[5]

Sylvania had a different perspective. For years they had been restive under the quota restrictions. They had tried to expand their market share by buying up other B licensees, but had not been able to expand beyond 5 percent of the market. The introduction of fluorescent lamps offered Sylvania an alternative way to grab a larger share of the market. Sylvania had been experimenting with fluorescence since 1931. When General Electric produced their first commercial fluorescent lamps in 1938, Sylvania faced a choice. General Electric offered Sylvania a B license with the same quota. But Sylvania chose the independent route, and rapidly developed their own fluorescent lamp (and their lawyers developed an independent patent position). Sylvania proceeded to market fluorescent lamps aggressively. Over the first five years of fluorescent lamp production, Sylvania increased its market share from under 5 percent for incandescent to 20 percent for fluorescent.[6] General Electric and Westinghouse were forced to follow suit, and sales of fluorescent lamps burgeoned.

The production of fluorescent lamps proceeded in three steps: the preparation of the phosphors, the coating of the tubes, and the finishing of the tubes. First, oxides of beryllium, zinc, magnesium, and manganese were mixed with silicic acid, milled to the specified particle size and fired to high temperatures. This process gave rise to the zinc beryllium silicate that was the most important phosphor, making up 13 percent of the mixture in the early Sylvania fluorescent tubes. Next, the phosphors were dissolved in lacquer solvents and flushed through the glass tubes. After the tubes were dried, they were baked to drive off the solvents. The ends of the tubes were wiped or brushed to provide powder-free bands onto which the end mounts could be sealed.[7] Since no reliable, quantitative air samples were measured, we cannot be sure where the exposures were most severe, but there do seem to have been significant amounts of phosphor dust in all the manufacturing areas of the plant . . . and soon there was disease.

THE EPIDEMIC IN SALEM

The first commercial fluorescent lamps in the United States were introduced at the World's Fair of 1939. By 1941, business was so good that Sylvania was building a new $500,000 production plant at Danvers. But until that plant was ready, Sylvania continued to manufacture fluorescent lamps at its old plant on Boston Street in Salem.[8] In 1941, in Building B of the old Boston Street plant, an epidemic led to the recognition of chronic beryllium disease. Late in 1941 and early in 1942, two of the women who worked in Building B became short of breath. At first, it was felt that they had tuberculosis, which was still an important public health problem. But the investigation at the Essex County Sanitarium revealed that neither of them was tuberculous. Rather, they both appeared to be suffering from a much less common disease, sarcoidosis. Now, it happened that Dr. Olin S. Pettingill, the Director of the Essex County Sanitarium, where these two women were cared for, was a close friend and sailing mate of Frank Healy, the Sylvania Vice President in charge of the Boston Street plant. Furthermore, Mr. Healy was a close friend of Dr. Thomas L. Shipman, the plant physician at the General Electric River Works facility. Dr. Shipman, in turn, was a friend of Manfred Bowditch, who, as director of the Massachusetts Division of Occupational Hygiene, was to head up the study of this epidemic.

Bowditch was the son of an old and important Massachusetts family. Although he was not a doctor, he eventually became involved in public health, first as an industrial hygienist at General Electric. It was there that he became a close friend of Thomas L. Shipman. This friendship continued even after Bowditch left General Electric in 1933 to become the first director of the State Division of Occupational Hygiene. For years afterward, the two of them would lecture jointly on occupational health to various industrial groups. Bowditch felt that "the General Electric Company has a medical department which is second to none."[9] Hardly a month went by without Bowditch hearing from Shipman about some professional concern. In 1943, Shipman wrote the Massachusetts House Committee on Ways and Means to support the budgetary allotment to Bowditch's Division.[10]

The Division of Occupational Hygiene first officially heard about the outbreak of sarcoid at the Boston Street plant in August 1942. Dr. Irving Tabershaw was at the plant to investigate silica exposures when Russell Tirrell, Sylvania's safety engineer, mentioned the two cases of sarcoid. Apparently, Shipman had suggested that the state agency would be helpful.[11] On September 9, 1942, Tirrell wrote directly to Bowditch describing the cases of sarcoid. Sylvania was "a little concerned to find that they both have this disease which is apparently rare, and not due to industrial exposure."[12] Tirrell agreed with Tabershaw that a cross-sectional X-ray study of Sylvania's employees was in order. A meeting was arranged with Dr. Alton S. Pope, the State Deputy Commissioner of Public Health, to plan an X-ray screening largely at government expense. However, negotiations broke down when Sylvania insisted on access to individual X-ray results and Dr. Pope insisted on the medical confidentiality of such results. Sylvania had changed its position:

> We do not feel that the hazards which we have in our plants warrant x-raying all our employees at the present time. If we had felt that the two cases we had were more or less due to industrial conditions, we might feel differently about this.[13]

Although Sylvania was convinced that the sarcoid was not related to any industrial exposure, one of the first patients disagreed and told Dr. Tabershaw that she was suing for compensation, ascribing her disease to mercury poisoning.[14] Thus, from the very beginning, Sylvania was concerned about workers' health, but more concerned about potential compensation liability.

Faced with Sylvania's ambivalence, the Division of Occupational Hygiene turned to Sylvania's largest competitor, General Electric. Tabershaw wrote to Walter Burchett at the General Electric Company Research Department in Nela Park, near Cleveland. Burchett responded that General Electric had not had any occupational disease among fluorescent lamp workers, but that silica, solvents, and mercury were potential hazards. This, too, set a pattern for the future; whenever Sylvania was not cooperating, the Division turned to General Electric.[15]

In December of 1942, one of the first two cases from Boston Street died. Her autopsy was performed by Dr. Timothy Leary, the Suffolk County Medical Examiner, and was witnessed by Dr. Harold L. Higgins, the former Chief of Pediatrics at the Massachusetts General Hospital, who was employed at that time by the Arrow Mutual Insurance Company, Sylvania's compensation carrier. Dr. Leary commented in his autopsy report:

> From the character and distribution of the lesions it is evident that we are dealing with a disease in the nature of an infection rather than with the effects of any toxic agent. . . . It is my opinion that M.K. came to her death from sarcoidosis, evidently an infectious disease. We know nothing of its method of infection but the evident close relation to tuberculosis suggests similar methods of transmission. From our present viewpoint it is difficult to see how or why sarcoidosis should be more prevalent in the electric than in other industries.[16]

Evidently, Dr. Leary felt strongly that Sylvania should not have to pay compensation in this case.

By January 15, 1943, when the Division of Occupational Hygiene held a conference of interested parties, several new cases had arisen. Sylvania now became more concerned. They purchased their own X-ray machine in order to undertake a cross-sectional screen of their workers. Moreover, they sent Dr. Higgins and Mr. Tirrell to visit their Towanda plant, which manufactured the fluorescent powders used in Salem. There they discovered three cases suspiciously like those in Salem, among workers in the sifting rooms. "Dr. Higgins described very dusty conditions in the Towanda plant in all departments. He stated that the most likely etiologic factor in his opinion is the beryllium oxide."[17] Over the next few months, more disease appeared. Three cases of a more acute chemical pneumonitis were reported from the Clifton Products beryllium refinery in Paynesville, Ohio. Bowditch sent his other physician, Dr. G. E. Morris, to check with the American Mutual Life Insurance Company; there Morris turned up "ten open cases of beryllium poisoning." "After reading these cases," Morris concluded, "I am convinced that the cases at Sylvania Electric are duplicates and that we should look for beryllium in the air at this plant."[18] On May 5, 1943, Morris wrote to Bowditch:

> It occurs to me that we should do something about ventilating or completely enclosing, if possible, all places where beryllium is handled at Sylvania without further delay.[19]

While the clinical and epidemiological evidence was pointing more and more directly to beryllium, the U.S. Public Health Service, in NIH Bulletin 181, *The Toxicology of Beryllium*, gave the metal a clean bill of health. This investigation was prompted by a series of articles from Italy, Germany, and the Soviet Union which described acute and chronic respiratory diseases from industrial exposures to beryllium.[20]

In view of these "reputedly poisonous properties of beryllium,"[21] this investigation proposed to measure the toxicity of beryllium and set maximum permissible exposure levels for workers "so far as is feasible by means of animal experiments."[22] These experiments consisted of exposing rats and guinea pigs to various beryllium compounds by intraperitoneal injection, by ingestion, and by inhalation. The study concluded that beryllium "of itself" is not toxic.[23]

> Since no particular toxicity was established for beryllium, it appears that whatever toxicity has been found to occur with the beryllium salts is due to the toxicity of the acid radical such as the fluoride or oxyfluoxide.[24]

On May 26, 1943, a second conference was held to discuss the "sarcoid" cases. This conference, unlike the first one in January, was national in scope. The meeting in Rochester, New York, had representatives from the Trudeau Institute, the U.S. Army Division of Occupational Hygiene, and the University of Pennsylvania. Dr. Tabershaw reviewed the data. He pointed out that the cluster of cases in a single plant suggested an industrial etiology. Although the specific cause was still in doubt, he suggested beryllium on the basis of epidemiologic evidence. Indeed, beryllium was present in the fluorescent powders, and a syndrome that seemed like "an acute stage of a similar disease" was occurring in beryllium extraction plants in Ohio and Pennsylvania. However, he conceded that the animal studies had so far shown beryllium silicate to be non-toxic. Dr. Leroy Upson Gardner, Director of the Trudeau Foundation, summarized the discussions more cautiously. He felt that the cluster was "more than a coincidence and perhaps we are dealing with a condition of industrial origin." He suggested a new series of animal studies and a chest X-ray screening of the industry.[25]

Gardner was an expert on occupational lung disease. When his career was interrupted by tuberculosis in 1918, he went to the Trudeau Sanitarium, where within a year he was well enough to join the staff of the Trudeau Foundation as a research pathologist.[26] As a patient at Saranac, Gardner acquired "a paternal feeling for the tuberculous workmen."[27] From this paternalism arose his long line of research into the pneumoconioses, the dust diseases. He started with silicosis. Using guinea pigs, he was able to show that the free silica content of dust correlated with silicosis at autopsy. Then, using tuberculous guinea pigs, he was able to show the synergistic effect of silica on the tubercle bacillus. However, having defined this synergism, he went on to argue that other irritant dusts and gases had no effect on tuberculosis, which therefore should not be a compensable disease. He was a man who felt torn between his paternal concerns for workers and his contractual obligations to industry, which funded most of his research. Just before his death, he told Harriet Hardy of the anguish it caused him not to publish his latest results on asbestosis in cats—results based on research he had done under contract to Johns-Manville.[28] Throughout his life he became "more and more associated with industry."[29] On his *curriculum vitae*, he declared he was a Republican; he was comfortable working with industrialists.

Gardner became involved in the "sarcoid" problem through his friendship with Manfred Bowditch. In June of 1943, Bowditch met with E. Finley Carter, the head of Industrial Relations for Sylvania, to discuss the future direction of the investigation. When Carter raised the question of a Medical Director to head the company's research effort, Bowditch recommended Gardner. By the end of June, Sylvania had hired Gardner as a consultant, paying him $5,000 and implying that they would raise a similar amount from their insurance carrier.[30] For the next three years, the investigation of the Salem "sarcoid" problem was directed by Manfred Bowditch, working for Massachusetts State Division of Occupational Hygiene, and his colleague, Dr. Leroy Gardner. As the months passed, relations between Sylvania and the researchers became strained, but Sylvania continued to fund Dr. Gardner.

A PAPER NOT PUBLISHED

The summer of 1943 was a confusing one for Gardner and Bowditch. The initial evidence had pointed to beryllium as the cause of a new occupational disease. By September 1, eight cases of "sarcoid" had been found among the workers at Sylvania's Boston Street plant. It was clear that something related to that plant caused the disease. Beryllium, as the only new substance in the process, was suspect. Moreover, there were clusters of acute pulmonary disease in three separate beryllium extraction plants using two different extraction processes. On August 20, Gardner wrote to Bowditch that "the element beryllium is the only common link and yet Fairhall (USPHS) tells us that his animal experiments lead him to believe that this is not of itself toxic."[31] Clinically and pathologically, the acute disease seemed different from the chronic disease, but as Gardner pointed out, "the beryllium workers themselves have potentially a heavy exposure and develop an acute disease rapidly whereas the lamp workers may have a very light exposure to similar substances producing a chronic disease slowly."[32] Three days after Shipman summarized Gardner's argument, Gardner was shifting position. On August 28 he wrote a letter to Carter claiming, "I am beginning to suspect that the pulmonary conditions in the beryllium manufacturing plants and your lamp plants are separate and distinct entities."[33] He argued that the pathological changes in the two cases were as different as "day from night."[34] Perhaps, he suggested, it was only coincidence that the two diseases occurred at the same time.

Dr. Joseph Aub, the lead expert at the Massachusetts General Hospital, agreed with Gardner's first opinion. Based on the clinical presentation of the disease, he felt "that the disease had two phases, the acute dyspnea due to bronchial irritation, which later subsides, and is followed by a more chronic reaction in some cases."[35] He suggested the need for more clinical studies, in particular blood and urine beryllium levels.

On September 22, the picture became more confusing. Fairhall wrote from Washington that he had found no beryllium in the lung tissue of M.K., the first woman to die in the Salem epidemic.[36] Although Fairhall had been the principal investigator of the USPHS animal studies which had proclaimed beryllium non-toxic "of itself," neither Bowditch nor Gardner doubted his results. (In retrospect, it turned out that Fairhall's analytic methods were inadequate to detect the beryllium in the lung sample.)

By now the situation was very confusing. The epidemiologic and clinical evidence pointed toward a biphasic disease caused by occupational exposure to beryllium. But the USPHS animal studies had been interpreted as negative, and the USPHS was unable to find beryllium in the lung tissue of the first dead worker in Salem. All involved in the investigation believed that they were dealing with an occupational disease. Whether or not they believed beryllium was the culprit depended on how they weighted the epidemiological versus the pathological evidence.

Although there was not yet agreement on the cause of the Salem epidemic, Bowditch and his colleagues agreed that it was something within the Boston Street Sylvania Plant. Realizing that this was a new occupational disease, Bowditch was anxious to get credit for its discovery. During the first week in October, he and Tabershaw wrote a brief paper with the modest title, *Pulmonary Sarcoidosis: A New Occupational Disease?* The paper took an epidemiologic perspective. Sarcoid, they argued, was an uncommon disease: only 6 cases of sarcoid were diagnosed among 43,620 admissions to Boston City Hospital in 1941. Thus, they were struck by the 8 cases they had discovered at a single fluorescent lamp factory with fewer than 1,000 employees.* As to the nature of the toxin, they remained agnostic, but stated that the evidence "seems to be against beryllium" when they wrote. Case histories of three of the eight were appended to the paper.[37] Bowditch sent a copy of the paper to Gardner, noting that the case histories were separate from the text because he was concerned that Sylvania might object to their inclusion in the paper.[38] Two days later Gardner responded, enthusiastically giving his complete approval.[39]

Tabershaw sent a copy of the manuscript to Sylvania asking for their comments. On October 26, Mr. Tirrell replied for the company. "We feel that this article is very much premature and that it also gives an indication that this disease may be of occupational origin."[40] He went on to explain that Dr. Gardner had suggested that the article could be modified to leave out "the reference that it (the disease) may be occupational from the manufacture of incandescent or fluorescent lamps."[41] Sylvania would be glad to go over the manuscript again after the revisions had been made. Tabershaw replied indignantly, "I do not think the article is in any way premature. The condition has been known to exist for well over a year. . . . If the suggestion was left out of the paper that the disease is occupational in origin, it would not be a modification but a complete emasculation . . . I am making these statements as a physician. The question of publication, however, is a matter which rests with the Director of this Division (Bowditch)."[42]

Further correspondence between Bowditch and Gardner indicates that Gardner had spoken with Tirrell in an attempt to arrange publication. A proposed linguistic compromise, which in place of "definitely stating" that the disease was occupational in origin merely "raised" the question "whether it may be occupational in origin," was not successful. Although Bowditch and Gardner do not discuss their final decision in the letters in the Division of Occupational Hygiene (DOH) archives, the article was never published.[43] A close reading of the correspondence between Bowditch and Gardner reveals no new developments which would have made the article obsolete. It is hard to escape the conclusion that it was suppressed because of Sylvania's objections.

THE COMPENSATION CASES

Sylvania's objections can be understood if we consider the question of compensation liability. From the very first cases, the workers felt that the disease came from their work. Thus, Sylvania did not want anything in print which would call the disease occupational. In February of 1944, Sylvania asked Gardner to testify in two pending compensation cases. In a difficult spot, he wrote to Bowditch, "Frankly, I do not know how to reply. Although I have strong suspicions, I still have no definite proof that their disease is occupational in origin nor do I know what to call it."[44]

---

*They pointed out that a survey of selective service records ruled out the possibility of a community-based epidemic.

Bowditch also wished not to testify at the compensation hearings, but for different reasons. The State Division of Occupational Hygiene had a legal right of entry to factories in Massachusetts, but it had no enforcement powers. Thus, to have an impact on health and safety, it depended on the cooperation of employers. In order to maintain a "reputation for impartial fairness," the DOH requested that none of its reports be used as evidence in law suits or workers' compensation suits.[45] In its advisory capacity to industries, it was very important for the DOH to maintain good relations with employers.[46]

Sylvania had two problems arising from the compensation hearings that took place in November and December of 1944. First, they were concerned about their legal liability. Equally important was their concern about adverse publicity. A handwritten note in the DOH files indicated that Sylvania's Russell Tirrell felt that the claims would be denied, but that if they were upheld, Sylvania would stop using the powders.[47]

Beginning on November 28, these and other front-page headlines appeared in the Boston papers: "Poison in Lamp Bulbs is Alleged," "Lamp Powder Blamed in Two Girls' Deaths," "Bulb Dust Victim Sobs at Hearing," "Doctors Testify Girls Made Ill in Lamp Plant," "Dr. Pettingill Says Powder Killed Girls," "Lamp Explodes at Poison Dust Hearings," "Dr. Leary Sees Nothing Toxic," and "No Cure for Bulb Poison." Gardner's cautious testimony that there was not enough evidence to know what caused the disease was lost in the barrage of headlines. The publicity could hardly have been worse.[48]

In response to the publicity, the officers of the United Electrical Workers Union that represented Sylvania's workers approached Dr. Pope, the Deputy Head of the State Department of Public Health, requesting an investigation. The State's response tells us a great deal about its position in labor-management disputes. After speaking with Dr. Pope, Bowditch "followed this up with Mr. R. C. Tirrell of the Sylvania Company who said that he saw no useful purpose to be accomplished by starting another and separate investigation of the "sarcoid" problem already the subject of so much work by this office and that of Dr. Gardner, but saw no objection to a statement by a suitable department of the State which might be reassuring to the Sylvania workers relative to health conditions under which fluorescent lamps are now being made."[49] Pope, in turn, wrote to the union explaining why there was "no undue risk . . . to the workers engaged in fluorescent lamp production."[50] Here we see the state agencies consulting Sylvania before taking any action and then issuing a statement to reassure the workers. Sylvania was viewed as an insider, a part of the investigation, while the union was seen as an outside intruder that needed to be pacified.

A month later, Sylvania settled out of court. Their insurance agent agreed to pay the workers as if they had won the cases, on the condition that the cases be withdrawn without a ruling.[51]

HARRIET HARDY AT THE DIVISION OF OCCUPATIONAL HYGIENE

In the fall of 1945, Dr. Harriet L. Hardy was hired to succeed Dr. Tabershaw at the DOH. As a physician, Hardy was unusual in that she was a woman and a liberal. Both her father and grandfather had been lawyers. Her mother was a suffragette, "an activist type, who supported various causes."[52] Her family were strongly in favor of the League of Nations, and Harriet had given speeches on "international friendship" as an undergraduate at Welles- ley. After completing her medical studies at Cornell and her internship at Philadelphia General Hospital, Dr. Hardy went to work as a physician at a private school in Western Massachusetts. Her social conscience was re- flected in her opening a clinic for the children of the poor rural families. She did not shy away from the opposition of local physicians afraid of losing customers. She returned to Boston as physician at Radcliffe College, but tired of caring for healthy young women and began to look for new chal- lenges in "clinical preventive medicine." Through Dr. Joseph Aub of the Massachusetts General Hospital, she got her job at the DOH.[53]

Her first day on the job, Bowditch presented Hardy with the voluminous and growing file on "Sylvania Sarcoid" and instructed her to devote all her time to it. Like a good epidemiologist, Dr. Hardy continued Bowditch's case finding. She reviewed the old cases and interviewed the new ones as they appeared. Her detective work led her to tuberculosis sanitaria, to hospital record rooms, and to fluorescent lamp factories. By May of 1946, Dr. Hardy had traced seventeen cases and was ready to publish. She wrote a paper for presentation to the Massachusetts Medical Society. But again Sylvania ob- jected. Dr. Hardy recounted the story:

Well, it proved that the insurance company that took care of the fluorescent lamp company had helped elect the Secretary of Labor for Massachusetts at that time and he felt indebted to them. They felt they must try to get my paper stopped from being published . . . and their doctor went to the medical society and tried to get the paper taken off the program. He went so far as to come down to Mr. Bowditch, my boss, and asked to see the paper and to change it as he saw fit. Mr. Bowditch conveniently was not there; he was out of town. So I talked to the doctor, whom I'd known in other connections. He said, "Harriet, you wouldn't want to make a mistake, would you?" By then my legal ancestors were beginning to rock around in my head, and I said, "Well, if I do make a mistake, it'll be my personal mistake." He said, "Well, if you let me have the paper, I'll just edit it."

Then Commissioner Moriarity, the Head of the Labor (Department) at that time . . . called me in and said, "Dr. Hardy, is there any reason why you shouldn't give that paper to the insurance company and let them go over it?" I said, "Commissioner Moriarity, if you ask me to change one dotting of an 'i' or crossing of a 't', what I will do is to resign and tell the newspapers exactly what happened." He said, "Now don't get so excited." I said, "I will. My father and grandfather were lawyers, my grandfather was a judge"—I went on like this. I said, "You can have a copy of the paper if you want to, but I am going to read the original one next week at the Society meeting."[54]

Dr. Hardy presented the original paper to a packed meeting of the Massachusetts Medical Society with the front rows filled with company officials, reporters, and a court stenographer who took down every word.

In many ways, Hardy's paper was similar to the one Tabershaw and Bowditch had written two and one-half years earlier. Entitled "Delayed Chemical Pneumonitis Occurring in Workers Exposed to Beryllium," it was modest in its claims.[55] In particular, it did not claim to show "the exact etiology" of the epidemic. Rather, it spoke of "a puzzling disease [that had] appeared in the employees of a concern manufacturing fluorescent lamps."[56] It mentioned Gardner's animal studies, which were still inconclusive. Then, after briefly summarizing the published literature, it described the seventeen cases including two autopsy reports. Tables summarized the clinical and laboratory findings. In closing, it suggested that the disease occurring in 17 workers in a single building "points to a common exposure" and that "evidence from the literature suggests that in some unknown way the fluorescent powders which contain beryllium compounds are of etiologic importance."[57] The publicity surrounding this paper made Harriet Hardy a celebrity of sorts; even today it is the best known of her many papers.

CONCLUSIONS

We are left then with the question of why Harriet Hardy published her paper while Manfred Bowditch withheld his earlier one. One possible explanation would be that new evidence had been uncovered since 1943 which made the paper more worthy of publication. However, an examination of the facts shows that this was not the case. It is true that by 1946 there were seventeen cases and six fatalities. In 1943, there had been only eight cases and one fatality. This makes the argument somewhat stronger, but even in 1943 the rate of disease in Building B was strikingly elevated. Moreover, the two major scientific objections to the earlier paper remained: there was still no animal model, and the autopsied lung tissue did not reveal beryllium.

A second explanation might look to the professional ideology of the actors. Robbins and Johnston have argued that scientific controversies are sometimes most bitter when the participants represent different disciplines, each with its own canons of evidence.[58] Certainly the epidemic in Salem attracted researchers from varied disciplines. Bowditch behaved like an epidemiologist, tracking down the cases and trying to find a common link between them. Trained as a pathologist, Gardner attempted to create an animal model and to find beryllium in the lungs of victims. For him, pathologic evidence carried the most weight. As a clinician who treated six cases of Salem sarcoid, Aub concentrated on describing the clinical features of the disease, and repeatedly suggested laboratory tests to find a clinical marker of exposure or disease. Hardy combined a clinical outlook similar to Aub's with epidemiologic detective work like Bowditch's. One might argue that pathologists would give more weight to the lack of an animal model and the absence of beryllium in autopsy material. But this can not explain what happened for two reasons. Neither of the authors of the earlier suppressed article were pathologists, and Gardner, the only pathologist of the group, was originally enthusiastic about the 1943 article.

A third explanation might lie in a diminution of the opposition to publication. But the circumstances surrounding Hardy's presentation of her paper belie this.

Finally, we must turn to the personal and political styles of the participants. Both Gardner and Bowditch saw themselves as consultants or advisors to industry. They were both concerned with maintaining their industrial connections. Gardner was always aware that he was a paid consultant for Sylvania.[59] For his part, Bowditch was concerned that the DOH maintain its "impartiality" so that it could be of service to industry. Both Bowditch and Gardner felt socially comfortable among industrialists. By way of contrast, Bowditch wanted nothing to do with the United Electrical Workers. His approach was to reassure them and get rid of them. In addition to feeling more comfortable relating to management than to labor, Bowditch seemed anxious to avoid controversy.

In contrast, Hardy was a feisty young lady. She did not draw back from controversy, and she viewed herself as a liberal champion of "causes."[60] Thus, in 1947, when the United Electrical Workers asked her to appear at their national convention, she readily agreed, although friends told her that the union was under attack as a subversive organization.[61] In her confrontation with the Arrow Mutual Insurance Company, she seemed to relish the fight. During her tenure at the DOH, she viewed her role as one of getting at the truth. When she felt she was not getting the whole story from management she would wander off to talk to workers. "I used to be pretty slick about losing the bossman."[62] When Bowditch felt that he was not getting the whole story from Sylvania, his instincts were to ask his friends at the General Electric Company.

Hardy and Bowditch had different views about the role of occupational health professionals, about the relative importance of their relations with labor and management, and about the way to resolve conflicts that arise. These differences explain why Bowditch did not publish his 1943 paper, while Hardy insisted on publishing hers in 1946.

NOTES

1. Thomas Kuhn, *The Structure of Scientific Revolutions*, (Chicago: University of Chicago Press, 1962).

2. Arthur A. Bright, Jr. and W. Rupert MacLaurin, "Economic Factors Influencing the Development and Introduction of the Fluorescent Lamp, *Journal of Political Economy* 51 (1943):432.

3. Ibid., pp. 436–38, 387.

4. Ibid., p. 435.

5. Ibid., p. 40.

6. Ibid., p. 439.

7. Emil C. Beyer, "The Manufacture of Fluorescent Lamps," in *Pneumoconiosis*, ed. Arthur J. Vorwald (New York: Paul B. Hoeber, Inc., 1950), pp. 30–32.

8. A. J. Breslin, "Exposures and Patterns of Disease in the Beryllium Industry," in *Beryllium: Its Industrial Hygiene Aspects*, ed. H. E. Stokinger (New York: Academic Press, 1966), pp. 77–78.

9. Commonwealth of Massachusetts, Division of Occupational Hygiene, *Publications*, vol. 5, number 193 (1941), Lecture 14.

10. Commonwealth of Massachusetts, Division of Occupational Hygiene, "Copies of 'BP' Letters."

11. Division of Occupational Hygiene 4912, letter from Skinner and Tabershaw to Bowditch, August 28, 1942. (All documents will be cited according to Division file number.)

12. Division of Occupational Hygiene 4912, letter from Tirrell to Bowditch, September 9, 1942.

13. Division of Occupational Hygiene 4912, letter from Tirrell to Bowditch, October 16, 1942; also letters from Bowditch to Tirrell, September 16, 1942, and Tabershaw to Bowditch, October 8, 1942.

14. Division of Occupational Hygiene 4912, letter from Tabershaw to Bowditch, October 13, 1942.

15. Division of Occupational Hygiene 4912, letters from Tabershaw to Burchett, November 30, 1942 and Burchett to Tabershaw, December 21, 1942.

16. Division of Occupational Hygiene 5662.

17. Division of Occupational Hygiene 6049, letter from Tabershaw to Bowditch, March 31, 1943.

18. Division of Occupational Hygiene 5662, letter from G. E. Morris to Bowditch, April 15, 1943.

19. Division of Occupational Hygiene 5662.

20. M. Berkovits and B. Izrael, "Lungen veranderungen bei der Intoxikation mit Fluor-beryllium," *Klinische Medizin* 18 (1940):117; S. Marradi Fabroni, "Patologia polmonare da polveri di berillio," *Medi de Lavore* 26 (1935):297; I. Gelman, "Poisoning by vapors of beryllium oxyfluorine," *Journal of Industrial Hygiene and Toxicology* 18 (1936):371; H. H. Weber and W. E. Engelhardt, "Untersuchung von Staubén aus der Berylliumgewinnung," *Zentralblat Gewerbehygiene und Unfallverhutung* 10 (1933):41.

21. Frances Hyslop et al., "The Toxicology of Beryllium," U.S. Public Health Service, National Institute of Health Bulletin No. 181, 1943, p. 4.

22. Ibid., p. 1.

23. Ibid., p. 48.

24. Ibid., p. 49.

25. Division of Occupational Hygiene 6049, Sarcoid Conference.

26. A. J. Lanza et al., "Tributes to Dr. Leroy Upson Gardner," *Occupational Medicine*, (July 4, 1947):1–16.

27. Ibid., p. 11.

28. Personal Communication, Dr. Harriet L. Hardy, April 17, 1982.

29. Lanza et al., p. 1.

30. Division of Occupational Hygiene 5662, Summary of Sarcoid Conference, June 14, 1943; also letters from Leroy Gardner to E. Finley Carter, June 20, 1943 and to Manfred Bowditch, June 20, 1943.

31. Division of Occupational Hygiene 6049, letter from Gardner to Bowditch, August 20, 1943.

32. Division of Occupational Hygiene 6183, letter from Shipman to Burchett, August 25, 1943.

33. Division of Occupational Hygiene 6049, letter from Gardner to Carter, August 28, 1943.

34. Ibid.

35. Division of Occupational Hygiene 6150, letter from Aub to Bowditch, September 3, 1943.

36. Division of Occupational Hygiene 6278, letter from Gardner to Bowditch, September 22, 1943.

37. Division of Occupational Hygiene 6535.

38. Division of Occupational Hygiene 6548, letter from Bowditch to Gardner, October 6, 1943.

39. Division of Occupational Hygiene 6535, letter from Gardner to Bowditch, October 8, 1943.

40. Division of Occupational Hygiene 6619, letter from Tirrell to Tabershaw, October, 26, 1943.

41. Ibid.

42. Division of Occupational Hygiene 6278, letter from Tabershaw to Tirrell, October 29, 1943.

43. Division of Occupational Hygiene 6619, letter from Bowditch to Gardner, November 2, 1943; Division of Occupational Hygiene 6460, letter from Gardner to Bowditch, November 19, 1943; Division of Occupational Hygiene 6492, letter from Bowditch to Gardner, November 23, 1943.

44. Division of Occupational Hygiene 6619, letter from Gardner to Bowditch, February 24, 1944.

45. Division of Occupational Hygiene 7886, letter from Bowditch to Bearl; Division of Occupational Hygiene 7584, letter from Bowditch to Gardner, November 10, 1944; Division of Occupational Hygiene 9332, letter from Philip Drinker and Bowditch to Polensvill, November 29, 1940.

46. Ibid.

47. Division of Occupational Hygiene 9332, dated December 8, 1944.

48. *Boston Post,* November 28, 1944; *Boston Record,* November 29, 1944; *Boston American,* December 1, 1944; Lynn Hem, November 29, 1944; *Boston Globe,* December 1, 1944; *Boston Herald,* December 2, 1944 and December 7, 1944; *Boston Daily Record,* December 2, 1944; and *Boston Daily Record,* December 20, 1944 for Gardner's testimony.

49. Division of Occupational Hygiene 9332, letter from Bowditch to Pope, December 29, 1944.

50. Division of Occupational Hygiene 9332, letter from Pope to Horie, UE, January 3, 1945.

51. Division of Occupational Hygiene 9332, letter from Bowditch to Gardner, January 16, 1945; letter from Tirrell to Bowditch, January 30, 1945; letter from Carter to Bowditch, February 5, 1945.

52. Personal Communication from Dr. Hardy, April 10, 1982.

53. Interview with Harriet Hardy, M.D., October 13–14, 1977, Oral History Project on Women in Medicine, Medical College of Pennsylvania, 1978.

54. Ibid., p. 39.

55. Harriet Hardy and Irving R. Tabershaw, "Delayed Chemical Pneumonitis Occurring in Workers Exposed to Beryllium Compounds," *Journal of Industrial Hygiene and Toxicology* 28 (1946):197–211.

56. Ibid., p. 197.

57. Ibid., p. 211.

58. David Robbins and Ron Johnston, "The Role of Cognitive and Occupational Differentiation in Scientific Controversies," *Social Studies of Science* 6 (1976): 345–368.

59. Division of Occupational Hygiene 7315, letters from Gardner to Leary, May 8, 1944, and from Gardner to Bowditch, May 8, 1944; Division of Occupational Hygiene 7315, letters from Bowditch to Gardner, August 7, 1944, and from Gardner to Bowditch, September 1, 1944.

60. Personal Communication, April 10, 1982.

61. *UE News*, October 4, 1947.

62. Personal Communication, April 10, 1982.

Part Three

Lead: Resistance to the Recognition of an Occupational and Environmental Poison

The development of the automobile industry spawned the growth of extensive state and interstate highway systems, allowing for the opening of new markets, a massive trucking industry, the American suburb, shopping malls, and even fast-food chains. These changes have provoked serious criticisms from consumer activists concerned with auto safety, urban planners worried about the strangling effects of suburbs on our great urban centers, and most recently environmentalists fearful that auto and truck emissions are disrupting the ecological balance and poisoning the air.

It has become clear in the last ten years that the health of the public is also threatened by the lead that has been added to gasoline for more than sixty years. Lead has been identified as a serious industrial poison for centuries. But it is only in the last decade that efforts to eliminate it from the environment have been even modestly successful. These three chapters address the history of earlier attempts to identify and control lead's impact on American society. Rosner and Markowitz detail the controversy over the introduction of lead into gasoline during the 1920s. They illustrate that public health concerns over lead were widely recognized and hotly debated. William Graebner follows with an analysis of industry's attempt to control research on leaded gasoline. He points to the evolving industry policy of control over scientific research as a central method by which regulation and popular understanding were thwarted. Ruth Heifetz looks at another aspect of lead in the environment, investigating the changing understanding of the effect of lead on human reproduction. She shows how the attention focused on female workers has obscured our understanding of the effects of these toxins on males' role in reproduction. The three articles raise important questions about the relationship between industry and public policy.

For much of our history, industry itself has done the research on new and potentially hazardous industrial substances. As these three articles indicate about just one substance, lead, even research carried out by seemingly objective agencies of government and universities is subject to widely different interpretations and, at various moments, manipulation.

CHAPTER **8**   *David Rosner and Gerald Markowitz*

"A Gift of God"?: The Public Health Controversy over Leaded Gasoline during the 1920s

A recent article in the *American Journal of Public Health* noted the high correlation between the lead content of soil in urban areas and the elevated blood-lead levels of children in these cities. An editorial in the same issue of the *Journal* suggested that the "use of leaded gasoline and [high] traffic density" helped explain this observation.[1] For most public health experts, the controversy over the possible adverse effects of leaded gasoline began in the 1970s. What we intend to show in this chapter is that as early as the 1920s public health experts, government officials, scientists, corporate leaders, labor, and the public were acutely aware of the dangers posed by the introduction of lead into gasoline. The depth of concern was manifested by the fact that leaded gasoline was banned in New York City for over three years, and in many other states and municipalities for shorter periods of time. In 1925, the production of leaded gasoline was halted for over nine months.

During the 1920s the petrochemical and auto industries emerged as the corporate backbone of the United States. Because the acceptance or rejection of leaded gasoline had profound implications for these industries, a spirited and often heated controversy arose. Public health professionals found themselves under intense pressure to sanction and minimize the hazards associated with the manufacture and use of this new potentially toxic substance, and the pages of the *American Journal of Public Health* were compromised during the months and years when the fate of leaded gasoline was decided. The debates of that era centered on issues of health and public policy that remain current today. Questions arose regarding the evaluation of health hazards associated with new and potentially harmful substances: How can scientists evaluate the relative importance of acute and chronic effects of toxic substances? What should constitute adequate proof of safety or harm? What business, professional, or government agencies should be responsible for evaluating possibly dangerous substances? How does one study potentially toxic substances while protecting the right to health of human subjects? Does industry have to prove a new substance safe, or do public health experts have to prove it dangerous? Because of scientific uncertainty concerning the safety or dangers posed by leaded gasoline, and because of the perceived need for this substance in the auto industry, the broader question became "what is the level of acceptable risk that society should be willing to assume for industrial progress?" What we will illustrate by examining this controversy is that at every stage of the debate the political, economic, and scientific issues were inextricably intertwined.

Before the 1920s the automobile industry was expanding and highly competitive. In addition to national manufacturers such as Ford, General Motors, and Studebaker, there were local companies, sometimes arising out of former bicycle manufacturers, that competed for special markets. Ford, however, dominated the pre-1920 market, producing nearly half of all the cars bought by Americans. Its Model T, small and cheaply produced, was the standard for the industry. In the 1920s General Motors developed a number of marketing and stylistic innovations that allowed it to replace Ford as the number one producer by the end of that decade. Alfred Sloan, president of General Motors, explained that their strategy called for creating demand "not for basic transportation, but for progress in new cars for comfort, convenience, power and style." Central to the creation of powerful and large automobiles was the development of a more efficient fuel capable of driving cars at greater speed. In 1922 Thomas Midgley and co-workers at the General Motors Research Laboratory in Dayton, Ohio discovered that when tetraethyl lead was introduced into gasoline it raised the compression, and hence the speed, by eliminating the "knock." This allowed for the development of the automobile essentially as we know it today.

General Motors, which had an interlocking directorship with the DuPont Chemical Company, quickly contracted with DuPont and Standard Oil of New Jersey to produce tetraethyl lead. It was immediately placed on sale in select markets on February 1, 1923. In 1924 DuPont and General Motors created the Ethyl Corporation to market and ultimately produce its final product. This was done in spite of the fact that industrial hygienists such as Alice Hamilton had long since identified lead as an industrial toxin.[2]

In the very year that Midgley and his co-workers at General Motors Research Corporation heralded the discovery of this powerful anti-knock compound, scientists inside and outside government warned that tetraethyl lead might be a potent threat to the public's health. William Mansfield Clark, a professor of chemistry, wrote to A. M. Stimson, Assistant Surgeon General at the Public Health Service, in October of 1922 warning of "a serious menace to the public health." He noted that in the early production of tetraethyl lead "several very serious cases of lead poisoning have resulted." He worried that its use in gasoline would result in environmental pollution, for "on busy thoroughfares it is highly probable that the lead oxide dust will remain in the lower stratum."[3]

Stimson believed that "the possibilities of a real health menace do exist in the use of such a fuel and it is deemed advisable that the Service be provided with some experimental evidence tending to support this opinion," and suggested that it was in the province of the Division of Chemistry and Pharmacology to conduct investigations of the dangers.[4] The Director of that division opposed this suggestion because such an investigation would take "a considerable period of time, perhaps a year," and that the results would be of little "practical use since the trial of the material under ordinary conditions [of use] should show whether there is a risk to man." He recommended instead that the Service depend upon industry itself to provide them with relevant data.[5]

A month later, H. S. Cumming, the Surgeon General, wrote to P. S. DuPont, chairman of the board of the DuPont Company, respectfully asking whether the public health effects of tetraethyl lead manufacturing and use had been taken into account. He was answered by Thomas Midgley himself, who acknowledged that the question "had been given very serious consideration . . . although no actual experimental data has been taken." Even without experimental data GM and DuPont were confident that "the average street will probably be so free from lead that it will be impossible to detect it or its absorption."[6]

DuPont and General Motors recognized that, in view of the general apprehension about the potential health hazards of tetraethyl lead, a purely private in-house study of its safety would be met by skepticism and rejection. Therefore, rather than conduct their own investigations, they worked out an agreement with the U.S. Bureau of Mines. The agreement stated that the General Motors Research Corporation would provide funding for an investigation of the dangers of tetraethyl lead, and that the Bureau of Mines would provide the facilities and the *imprimatur* of the U.S. Government on the results of such an investigation. GM, through its prime negotiator Charles Kettering, requested one other proviso: that "the Bureau refrain from giving out the usual press and progress reports during the course of the work, as [Kettering] feels that the newspapers are apt to give scare headlines and false impressions before we definitely know what the influence of the material will be."[7]

It was clear to everyone concerned that this was a politically explosive inquiry. Even the correspondence between scientists reflected its political nature. Bureau of Mines chief chemist S. C. Lind, for example, wrote to the Superintendent of the Pittsburgh Bureau of Mines Field Station where the investigation was being carried out, objecting to the government's use of the trade name "ethyl" when referring to tetraethyl lead gasoline. "Of course their [GM's] object in doing so are fairly clear, and among other things they are not particularly desirous of having the name 'lead' appear in this case. That is alright from the standpoint of the General Motors Company but it is quite a question in my mind as to whether the Bureau of Mines would be justified in adopting this name so early in the game before it has had the support of popular usage." Fieldner, the Superintendent, replied, however, that the avoidance of "the use of 'lead' in the interbureau correspondence" was intentional because of leaks to the newspapers. Since the Bureau had agreed to a blackout of information, he asserted that "if it should happen to get some publicity accidentally, it would not be so bad if the word 'lead' were omitted as this term is apt to prejudice somewhat against its use."[8]

The willingness of the Bureau of Mines to avoid publicity and even to avoid accurate scientific terminology in favor of a trade name reflected the tentativeness with which the Bureau approached the giant corporations, GM and DuPont. This can be seen in the sequence of agreements that would develop between this government agency and GM, DuPont, and the newly-created Ethyl Gasoline Corporation in the coming months—while the critical research into the health effects of tetraethyl lead progressed. The first agreement, in September 1923, between the General Motors Research Corporation and the Bureau allowed relative freedom for the Bureau to report its final conclusions.[9] However, by June 1924, General Motors sought much greater control over the final product. Not only did they demand that no publicity concerning the research be given to the popular press; now they added to their contract the stipulation that "all manuscripts, before publication, will be submitted to the Company for comment and criticism."[10] Two months after the Bureau acquiesced to this new stipulation, the newly created Ethyl Corporation asked that their proposed contract be modified in two respects. First, Ethyl requested that there be a dollar limit on the maximum expenses that the company would incur. Second, and most important, Ethyl asked "that before publication of any papers or articles by your Bureau, they should be submitted to them [Ethyl] for comment, criticism, and *approval*" (our emphasis). These changes were incorporated into the new contract, giving the Ethyl Corporation veto power over the research of the United States government.[11]

Despite the protestations of GM, DuPont, and the government that no information should be released before completion of the study, it is clear from the unpublished correspondence that they violated this agreement when it appeared that the preliminary results pointed toward a vindication of the companies' faith in tetraethyl lead. In July 1924, two years after tetraethyl lead was first put on the market in the Midwest and the East Coast, and five months before the preliminary report was released, GM's director of research, Graham Edgar, wrote to Dr. Paul Leech of the American Medical Association that the results of the Bureau of Mines' research would show "that there is no danger of acquiring lead poisoning through even prolonged exposure to exhaust gases of cars using Ethyl Gas." He further erroneously assured the AMA that "poisoning from carbon monoxide would arise long before the concentration of lead would reach a point where even cumulative poisoning is to be feared."[12]

The federal government's role was interpreted by many as a sign of apparent collusion with GM, DuPont, Standard Oil, and Ethyl to certify the safety of tetraethyl lead. Yandell Henderson of Yale University, a leading public health physiologist, wrote an angry letter to R. R. Sayers, the coordinator of the government's activities, pointedly rejecting an offer to have a role in the government's research. "As regards your suggestion that you might assign us [at Yale's Laboratory of Applied Physiology] a part in the investigation which you are carrying out for the General Motors on tetra-ethyl lead, I feel that I should want a greater degree of freedom of investigation and funding—in view of the immense public, sanitary, and industrial questions involved—than the subordinate relation which you suggest would allow. It seems to me extremely unfortunate that the experts of the United States Government should be carrying out this investigation on a grant from the General Motors." He continued that he felt "very strongly that there is the most urgent need for an absolutely unbiased investigation."[13] C. W. Deppe, the owner of a competing motor car company, was much blunter in his criticism of the government's relationship to GM: "May I be pardoned if I ask you frankly now, does the Bureau of Mines exist for the benefit of Ford and the G.M. Corporation and the Standard Oil Co. of New Jersey, and other oil companies party to the distribution of the Ethyl Lead Dopes, or is the Bureau supposed to be for the public benefit and in protection of life and health?"[14]

The dangers posed by the widespread introduction of leaded gasoline were brought to the public's attention by a disaster that occurred in the Standard Oil experimental labs in Elizabeth, New Jersey on October 26, 1924. Over the course of the following five days, five workers would die and thirty-five others would show severe neurological symptoms of organic lead poisoning. Of forty-nine workers in the tetraethyl lead processing plant, over 80 percent would be severely poisoned. In a page one story the next day, the *New York Times* headline reported "Odd Gas Kills One, Makes Four Insane." The *Times* quoted the company doctor, who suggested that "nothing ought to be said about this matter in the public interest," and one of the supervisors at the Bayway facility, who said "these men probably went insane because they worked too hard." The father of the dead man, however, "was bitter in denunciation of conditions at this plant." The father told reporters that "Ernest was told by the doctors at the plant that working in the laboratory wouldn't hurt him. Otherwise he would have quit. They said he'd have to get used to it."[15]

After this initial revelation, every major newspaper in New York began to report on conditions at the plant. Day after day, the *Times,* the *New York World,* and other newspapers revealed deaths and occupationally-related insanity due to what the newspapers called "looney gas."[16] The company continually sought to deny management's responsibility for the tragedy. Thomas Midgley appeared at a press conference and said that true responsibility for the crisis rested with the workers. He said that at another plant "the men, regardless of warnings and provision for their protection, had failed to appreciate the dangers of constant absorption of the fluid by their hands and arms." Midgley and other company representatives went even farther in their attempt to avoid responsibility for the fact that 40 of the 49 men who worked in the plant either died or went insane. In denying responsibility for the tragedy "officials of the company replied that the rejection of many men as physically unfit to engage in the work at the Bayway plant, daily physical examinations, constant admonitions as to wearing rubber gloves and using gas masks and not wearing away from the plant clothing worn during work hours should have been sufficient indication to every man in the plant that he was engaged 'in a man's undertaking.' " Despite Standard Oil's attempt to shift blame to workers, others were reaching different conclusions. The Union County prosecutor asserted that he was "satisfied many of the workers did not know the danger they were running. I also believe some of the workers were not masked nor told to wear rubber gloves and rubber boots."[17] The New Jersey Commissioner of Labor said he had never been informed that the workers in the Bayway plant were potentially in danger. "Secrecy surrounding the experiments was responsible for the Labor Department's lack of knowledge of them," an official said.[18] Under the relentless pressure of daily revelations and investigations, Standard Oil acknowledged, after the fifth victim had died, "that it was known that this gas had collected a previous toll of death and insanity before the 49 employees were exposed to it at the Elizabeth plant."[19]

These deaths and the continuing controversy stimulated a renewed concern about the potential public health dangers from the exhaust produced by leaded gasoline. Despite Standard Oil's assurance that no "perils existed in the use of this gas in automobiles," New York City, Philadelphia, and many other municipalities and states banned the sale of leaded gasoline.[20]

The day after the fifth and last victim died, and in the midst of growing public skepticism about this new chemical, the Bureau of Mines released its preliminary findings on the possible dangers of leaded gasoline to the general public. The *New York Times* headline summed up the report: NO PERIL TO PUBLIC SEEN IN ETHYL GAS—BUREAU OF MINES REPORTS AFTER LONG EXPERIMENTS WITH MOTOR EXHAUSTS—MORE DEATHS UNLIKELY. They also reported "the investigation carried out indicates the danger of sufficient lead accumulation in the streets through the discharging of scale from automobile motors to be seemingly remote." The report exonerated tetraethyl lead.[21] Despite the interest of the manufacturers to use the report to reassure the public, the circumstances of the workers' deaths only served to undermine the credibility of the Bureau of Mines' findings. Specific criticisms came from a number of different sources. Scientists and labor activists alike found fault with the report. E. E. Free, editor of the prestigious *Scientific American* magazine, was skeptical of R. R. Sayers's assurances that the Bureau of Mines could find no evidence of lead poisoning in the animals.[22] Cecil K. Drinker, editor of the *Journal of Industrial Hygiene* and professor of public health at Harvard, and Dr. David Edsall, Dean of the Harvard Medical School, were also critical. In early January of 1925, Drinker wrote to Sayers a pointed letter that concluded,

"As an investigation of an important problem in public health upon which a great deal of inexact data has already appeared, the report is inadequate.[23] Alice Hamilton concurred with Drinker's position and noted the "desirability of having an investigation made by a public body which will be beyond suspicion."[24]

Perhaps the strongest criticism of the Bureau of Mines' report came from the Workers' Health Bureau and one of its chief scientific advisors, Yandell Henderson of Yale University. Even before the report was issued, the Workers' Health Bureau, an organization of pro-labor activists devoted to investigating and organizing around occupational safety and health issues, wrote to one of their consultants, Emery Hayhurst, the noted industrial hygienist of the Ohio Department of Health, calling for a united stand to oppose lead in gasoline. They pointed out that the crisis at Bayway indicated that both workers and the general public were in danger of lead poisoning if lead were allowed to remain in gasoline. Hayhurst responded that he had no information that "tetraethyl lead is dangerous to the public nor to users or handlers of it in garages, filling stations, etc."[25] Henderson, upon whom the Workers' Health Bureau depended for much of their information about the dangers of tetraethyl lead, voiced the public health profession's nagging fear regarding the fact that "this investigation is financed by the Ethyl Gas Corporation" and that in spite of many protests "the investigators in the Bureau of Mines have used experimental conditions which are fundamentally unsuited to afford information on the real issues."[26]

This attack by scientists, public health experts, and activists on the quality and integrity of the report forced those who favored the introduction of lead into gasoline to begin a counteroffensive. Emery Hayhurst emerged as one of the key figures in the attempt to "sell" tetraethyl lead to the American public. Hayhurst was important in the following months and years because of his established reputation as a respected and independent industrial hygienist. But what was not known about Dr. Hayhurst during the months of struggle around this issue was that while he was advising organizations like the Workers' Health Bureau about industrial hygiene matters, he was also working for the Ethyl Corporation as a consultant.[27] Also, it is evident from correspondence between Hayhurst and the Public Health Service that Hayhurst was supplying advocates of tetraethyl lead with information regarding the tactics to be used by their opponents. Indeed, even before the Bureau of Mines had issued its report, Hayhurst had decided that tetraethyl lead was not an environmental toxin. He had advised the Bureau of Mines to include a statement that "the finished product, Ethyl Gasoline, as marketed and used both pure or diluted in gasoline retains none of the poisonous characteristics of the ingredients concerned in its manufacture and blending."[28] But even more damning was that, in another letter to Sayers when the attacks on the report were mounting, Hayhurst secretly sent to the Public Health Service criticisms that the Workers' Health Bureau had developed so that the Government could be prepared to reply. Although the Workers' Health Bureau had specifically refrained from sending these comments to the government, Hayhurst violated their trust.[29] Hayhurst and Sayers also worked together to build public and professional support for the Bureau of Mines' and the Ethyl Corporation's position that tetraethyl lead was not a public health danger. Sayers urged that Hayhurst counter the criticisms of Drinker and Edsall with a review or editorial of his own in support of the report. Hayhurst replied that he had prepared an editorial for the *American Journal of Public Health* and that the unsigned editorial proclaimed, "Observational evidence and reports to various health officials over the country . . . so far as we have been able to find out, corroborated the statement of 'complete safety' so far as the public health has been concerned."[30]

This propaganda effort was incapable of quelling the doubts about the safety of leaded gasoline or the integrity of the Bureau of Mines report. It also became apparent that the companies involved were engaging in a cover-up of other deaths and illnesses among their workers in other plants. In light of the publicity over Bayway, it was soon reported that other workers had died handling tetraethyl lead at the DuPont chemical plant at Deepwater, New Jersey and at the General Motors Research Division site in Dayton, Ohio. The Workers' Health Bureau, for instance, began to catalog the deaths and illnesses of workers at these plants, showing that since September of 1923 at least two men had died at Dayton and four more had died at Deepwater.[31] The *Times* later reported the difficulties that editors and reporters had in following the story. For example, the *Times* noted that there was nothing in the local paper about the death of Frank W. (Happy) Durr, who had worked for DuPont for 25 years. Durr had literally given his life to the company; he had begun working for DuPont as a child of twelve and died from exposure to tetraethyl lead at only 37. The editor of *The Record* told the *Times:* "I guess the reason we didn't print anything about Durr's death was because we couldn't get it. They [DuPont] suppress things about the lead plant at Deepwater. Whatever we print we pick up from the workers." The *Times* went on to describe the control that DuPont exercised over the local hospital to which its poisoned workers were sent, indicating that it was almost impossible to get information from the hospital about the source of the workers' problems. Despite this, the *Times* was able to uncover the fact that there were over 300 cases of lead poisoning among workers at the Deepwater plant during the previous two years. The workers at the DuPont facility knew that there was something amiss, and had dubbed the plant "the house of the Butterflies" because so many of their colleagues had hallucinations of insects during their bouts of lead poisoning: "The Victim pauses, perhaps while at work or in a rational conversation, gazes intently at space and snatches at something not there." The *Times* reported that "about 80% of all who worked in 'the House of the Butterflies,' or who went into it to make repairs were poisoned, some repeatedly."[32]

As a result of these continuing revelations and public disquiet over the Bureau of Mines report, the Surgeon General of the Public Health Service contemplated calling a national conference to assess the tetraethyl lead situation. In a frank letter from Haven Emerson, the eminent public health leader, to the Surgeon General, the concerns of public health officers were clearly spelled out. Far from accepting the Bureau of Mines' report as an aid for resolving the issue, he suggested that the report was having "a widespread, and to my mind harmful influence on public opinion and the actions of public agencies." He believed that it would be "well worthwhile to call those whom you intend to a conference promptly. . . . The impression is gaining way that the interests of those who may expect profit from the public sale of tetraethyl lead compounds have been influential in postponing such a meeting."[33] Despite some indication that R. R. Sayers opposed such a conference and may have delayed it,[34] the Surgeon General announced at the end of April 1925 that he was calling together experts from business, labor, and public health to assess the tetraethyl lead situation. In his call for the meeting, Surgeon General Hugh Cumming said that leaded gasoline "is a public health question of extreme seriousness . . . if this product is actually causing slow poisoning and serious effects of a cumulative character."[35] On May 20, 1925 the conference convened in Washington with every major party represented. At this conference the ideologies of the different participants were clearly and repeatedly laid out and provide an important forum in which we can evaluate the scientific, political, economic, and intellectual issues surrounding this controversy. In the words of one participant, the conference gathered together in one room "two diametrically opposed conceptions. The men engaged in industry, chemists, and engineers, take it as a matter of course that a little thing like industrial poisoning should not be allowed to stand in the way of a great industrial advance. On the other hand, the sanitary experts take it as a matter of course that the first consideration is the health of the people."[36]

The conference opened with statements from General Motors, DuPont, Standard Oil, and the Ethyl Corporation outlining the history of the development of leaded gasoline and the reasons why they believed its continued production was essential. In their discussions, three general themes emerged as central arguments of the companies. First, the manufacturers maintained that leaded gasoline was essential to the industrial progress of America. Second, they maintained that any innovation entails certain risks. Third, they stated that the major reason that deaths and illnesses occurred at their plants was that the men who worked with the materials were careless and did not follow instructions.

C. F. Kettering and Robert Kehoe both stressed the importance of tetraethyl lead as a means of conserving motor fuel. But Frank Howard, representing the Ethyl Gasoline Corporation, provided the most complete rationale for the continued use of tetraethyl lead in gasoline. He noted that it was not possible to abstract the questions of public health from broader economic and political issues. "You have but one problem," he remarked rhetorically, "Is this a public health hazard?" He answered that "unfortunately, our problem is not that simple." Rather he posited that automobiles and oil were central to the industrial progress of the nation, if not the world. "Our continued development of motor fuels is essential in our civilization," he proclaimed. Noting that at least a decade of research had gone into the effort to identify tetraethyl lead, he called its discovery an "apparent gift of God." By casting the issue in this way, Howard put the opposition on the defensive, making them appear to be reactionaries whose limited vision of the country's future could permanently retard progress and harm future generations. "What is our duty under the circumstances?" he asked. "Should we say, 'No, we will not use' " a material that is "a certain means of saving petroleum? Because some animals die and some do not die in some experiments, shall we give this thing up entirely?"[37]

The stark portrayal of tetraethyl lead as a key to the industrial future of the nation led naturally into industry's argument that any great advance required some sacrifice. Dr. H. C. Parmelee, editor of *Chemical and Metalurgical Engineering*, stated, "The research and development that produced tetraethyl lead were conceived in a fine spirit of industrial progress looking toward the conservation of gasoline and increased efficiency of internal combustion motors." Parmelee believed that the companies did their best to safeguard the workers. In the end, he said, "its casualties were negligible compared to human sacrifice in the development of many other industrial enterprises."[38]

The final part of the industries' position was that workers, rather than the companies, were at fault for the tragedies at Bayway, Deepwater, and Dayton. Acknowledging that there were "certain dangers" inherent in the production of this essential industrial product, the Standard Oil Company asserted that "every precaution was taken" by the company to protect their workers. Thomas Midgley, Jr., vice president of General Motors and known to many as "the Father of Ethyl Gas," was more pointed at the conference. He said that the lesson that the companies had learned from this whole experience was that "the essential thing necessary to safely handle [tetraethyl lead] was careful discipline of our men . . . [tetraethyl lead] becomes dangerous due to carelessness of the men in handling it." In an earlier statement to the *New York World*, Midgley explained what this discipline consisted of: "The minute a man shows signs of exhilaration he is laid off. If he spills the stuff on himself he is fired. Because he doesn't want to lose his job, he doesn't spill it." Midgley's own recklessness was revealed at a news conference in which he sought to downplay the toxicity of tetraethyl lead. When he was asked by a reporter if it was dangerous to spill the chemical on one's hands, Midgley dramatically "had an attendant bring in a quantity of pure tetraethyl." He then "washed his hands thoroughly in the fluid and dried them on his handkerchief. 'I'm not taking any chance whatever,' he said. 'Nor would I take any chance doing that every day.' " He did this act in spite of the fact that he had only a year before taken a prolonged vacation in Florida in order to cure himself of lead poisoning.[39]

Those who opposed the introduction of leaded gasoline disagreed with every fundamental position of industry representatives. First, opponents pointed out that what we would now denote as inorganic lead compounds were already known to be slow, cumulative poisons that should not be introduced into the general environment. Second, they believed that the Federal Government had to assume responsibility for protecting the health of the nation. Third, they rejected the notion that the workers were the ones responsible for their own poisoning. Fourth, and most important, because they believed that the public's health should take precedence over the needs of industry, they argued that the burden of proof should be on the companies to prove tetraethyl lead was *safe*, rather than on opponents to prove that it was dangerous.

Dr. Yandell Henderson, the Yale physiologist, was the strongest and most authoritative critic of industry. He told the conference that lead was a serious public health menace that could be equated to infectious diseases then affecting the nation's health. Henderson was horrified at the thought that hundreds of thousands of pounds of lead were going to be deposited in the streets of every major city of America. The problem was not that more people would die like those in Bayway or Deepwater, but that "the conditions would grow worse so gradually and the development of lead poisoning will come on so insidiously . . . that leaded gasoline will be in nearly universal use and large numbers of cars will have been sold . . . before the public and the government awaken to the situation."[40]

To meet such a public health menace, Henderson and other critics believed that it was essential that the federal government take an active role in controlling leaded gasoline. Unlike industry spokespeople who defined the problem as one of occupational health and maintained that individual vigilance on the part of workers could solve the problem, Henderson believed that leaded gasoline was a public health and environmental health issue that required federal action. Harriet Silverman of the Workers' Health Bureau amplified his point by directly attacking the idea put forward by industry that the workers were responsible for their own poisoning. "I ask you gentlemen to consider the fact that you are asked to allow a man to be subjected to contact with a poison which is considered hazardous by the leading scientists of the country. And when you expose them to the poison out of which the manufacturers are making profits, the manufacturers penalize those men by making them forfeit a day's wage."[41]

Opponents were most concerned, however, about the propaganda of the industry that equated the use of lead with industrial progress and the survival of our civilization itself. Reacting to the Ethyl Corporation representative's statement that tetraethyl lead was a "gift from God," Grace

Burnham of the Workers' Health Bureau said that she wanted "to make a statement that this . . . was not a gift of God when those 11 men were killed or those 149 were poisoned." She angrily questioned the priorities of "this age of speed and rush and efficiency and mechanics" and said that "the thing we are interested in the long run is not mechanics or machinery, but men." A. L. Berres, Secretary of the Metal Trades Department of the American Federation of Labor, also rejected the prevalent conception of the 1920s that "the business of America was business." He told the conference that the American Federation of Labor opposed the use of tetraethyl lead. "We feel that where the health and general welfare of humanity is concerned, we ought to step slowly." But it was Yandell Henderson who summarized the opponents' position in a private letter to R. R. Sayers of the Bureau of Mines: "In the past, the position taken by the authorities has been that nothing could be prohibited until it was proved to have killed a number of people. I trust that in the future, especially in a matter of this sort, the position will be that substances like tetraethyl lead can not be introduced for general use until it is proved harmless."[42]

For the vast majority of public health experts at the conference, the problem was how to reconcile the opposing views of those favoring industrial progress and those frightened by the potential for disaster. Although everyone hoped that science itself would provide an answer to this imponderable dilemma, the reality was that all evidence had thus far been ambiguous. One problem was that no one in the 1920s had a model for explaining the apparently idiosyncratic occurrence of lead poisoning. Even the medical director of Reconstruction Hospital in New York, probably the only facility devoted exclusively to the study and treatment of occupational disease and accidents, could not explain the strange manifestations of chronic tetraethyl lead poisoning. He said that of the 39 patients he treated after the Bayway disaster, "some of these individuals gave no physical evidence and no symptoms or any evidence that could be found by a physical examination that would indicate that they were ill, but at the same time showed lead in the stools." He concluded that "this suggested to me that perhaps a man may be poisoned from the tetraethyl lead without showing clinical evidence and that therefore, there may be a considerable number of individuals so poisoned who have not come under observation." The policy implication for him was that ethyl gasoline "should be withheld from public consumption until it is conclusively shown that it is not poisonous."[43]

The country's foremost authority on lead, Dr. Alice Hamilton, agreed with those opposed to tetraethyl lead. She told the conference that she believed the environmental health issues were more important than the occupational health and safety issues because she doubted that any effective measures could be implemented to protect the general public from the hazards of widespread use of leaded gasoline. "You may control conditions within a factory," she said, "but how are you going to control the whole country?" In a more extended commentary on the conference and the issues that it raised, Hamilton stated, "I am not one of those who believe that the use of this leaded gasoline can ever be made safe. No lead industry has ever, even under the strictest control, lost all its dangers. Where there is lead some case of lead poisoning sooner or later develops, even under the strictest supervision."[44]

Most public health professionals did not agree with Henderson and Hamilton. Many took the position that it was unfair to ban this new gasoline additive until definitive proof existed that it was a real danger. In the face of industry arguments that oil supplies were limited and that there was an extraordinary need to conserve fuel by making combustion more efficient, most public health workers believed that there should be overwhelming evidence that leaded gasoline actually harmed people before it was banned. Dr. Henry F. Vaughan, president of the American Public Health Association, said that such evidence did not exist. "Certainly in a study of the statistics in our large cities there is nothing which would warrant a health commissioner in saying that you could not sell ethyl gasoline," he pointed out. Vaughan acknowledged that there should be further tests and studies of the problem, but said that "so far as the present situation is concerned, as a health administrator I feel that it is entirely negative." Emery Hayhurst also argued this at the Surgeon General's Conference. He believed that the widespread use of leaded gasoline for 27 months "should have sufficed to bring out some mishaps and poisonings, suspected to have been caused by tetraethyl lead."[45]

While Hayhurst and others publicly voiced little concern over the dangers of leaded gasoline, in private they were not quite as assured. One investigator from Columbia University, Frederick Flinn, avoided voicing an opinion regarding the public health danger of leaded gasoline at the conference. However, in a personal communication to R. R. Sayers of the United States Public Health Service and the Bureau of Mines, he told of his fears. "The more I work with the material [tetraethyl lead] the more I am confused as to whether it is a real public health hazard," he began. He felt that much depended upon the special conditions of exposure in industry and on the street, but in the end stated that he was "convinced that there is some hazard—the extent of which must be studied around garages and filling stations over a period of time and by unprejudiced persons." Given the fact that Flinn did this study for the Ethyl Corporation, it is not surprising that he ended his letter by saying that "of course, you must understand that my remarks are confidential." Emery Hayhurst was even more candid in his private correspondence to Sayers. He told Sayers that he had just received a letter from Dr. Thompson of the Public Health Service saying that "lead has no business in the human body. . . . That everyone agrees lead is an undesirable hazard and the only way to control it is to stop its use by the general public." Hayhurst, however, acknowledged to Sayers that political and economic considerations influenced his scientific judgment.

Personally I can quite agree with Dr. Thompson's wholesome point of view, but still I am afraid human progress cannot go on under such restrictions and that where things can be handled safely by proper supervision and regulation they must be allowed to proceed if we are to survive among the nations. Dr. Thompson's arguments might also be applied to gasoline and to the thousand and one other poisons and hazards which characterize our modern civilization.[46]

Despite the widespread ambivalence on the part of public health professionals and the opposition to any curbs on production on the part of industry spokespeople, the public suspicion aroused by the proceeding year's events led to a significant victory for those who opposed the sale of leaded gasoline. The Ethyl Corporation announced at the end of the conference that it was suspending the production and distribution of leaded gasoline until the scientific and public health issues involved in its manufacture could be resolved. Also, the conference called upon the Surgeon General to organize a blue ribbon committee of the nation's foremost public health scientists to conduct a study of leaded gasoline. Among those asked to participate were David Edsall of Harvard University, Julius Steiglitz of the University of Chicago, and C.-E.A. Winslow of Yale University. For Alice Hamilton and other opponents of leaded gasoline, the conference appeared to be a major victory, for it took the power to decide on the future of an important industrial poison away from industry, placing it in the hands of university scientists. "To anyone who has followed the course of industrial medicine for as much as ten years," Alice Hamilton remarked one month after the conclusion of the conference, "this conference marks a great progress from the days when we used to meet the underlings of the great munition makers [during World War I] and coax and plead with them to put in the precautionary measures. . . . This time it was possible to bring together in the office of the Surgeon General the foremost men in industrial medicine and public health and the men who are in real authority in industry and to have a blaze of publicity turned on their deliberations."[47]

The initial euphoria over the apparent victory of "objective" science over political and economic self-interest was short-lived. The blue ribbon committee, mandated to deliver an early decision, designed a short-term, and in retrospect very limited, study of garage and filling station attendants and chauffeurs in Dayton and Cincinnati. The study consisted of following four groups of workers, 252 people in all. Of these, 36 men were controls employed by the city of Dayton as chauffeurs of cars using gasoline without lead, while 77 were chauffeurs who had been using leaded gasoline for two years. Also, 21 others were controls employed as garage workers or filling station attendants where unleaded gasoline was used, and 57 were engaged in similar work where tetraethyl gas was used. As another means of comparison, 61 men were tested in two industrial plants in which there was known to be serious exposure to lead dust. As a result of this study, the committee concluded seven months after the conference that "in its opinion there are at present no good grounds for prohibiting the use of ethyl gasoline . . . provided that its distribution and use are controlled by proper regulations," and suggested that the Surgeon General formulate specific regulations with enforcement by the states. Although it appears that the committee rushed to judgment in only seven months, it must be pointed out that this group saw their study as only interim, to be followed by longer-range studies in the coming years. In their final report to the Surgeon General, the committee warned that

> it remains possible that if the use of leaded gasoline becomes widespread conditions may arise very different from those studied by us which would render its use more of a hazard than would appear to be the case from this investigation. Longer experience may show that even such slight storage of lead as was observed in these studies may lead eventually in susceptible individuals to recognizable or to chronic degenerative diseases of a less obvious character.

Recognizing that their short-term retrospective investigation was incapable of detecting such danger, the committee concluded that further study by the Government was essential:

> In view of such possibilities the committee feels that the investigation begun under their direction must not be allowed to lapse. . . . It should be possible to follow closely the outcome of a more extended use of this fuel and to determine whether or not it may constitute a menace to the health of the general public after prolonged use or other conditions not now foreseen. . . . The vast increase in the number of automobiles throughout the country makes the study of all such questions a matter of real importance from the standpoint of public health and the committee urges strongly that a suitable appropriation be requested from Congress for the continuance of these investigations under the supervision of the Surgeon General of the Public Health Service.[48]

These suggestions were never carried out, and all future studies of the use of tetraethyl lead were conducted by the Ethyl Corporation and scientists employed by them. In direct contradiction to the recommendations of the committee, Robert Kehoe, who carried out the studies for Ethyl, wrote: "as it appeared from their investigation that there was no evidence of immediate danger to the public health, it was thought that these necessarily extensive studies should not be repeated at present, at public expense, but that they should be continued at the expense of the industry most concerned, subject, however, to the supervision of the Public Health Service." It should not be surprising that Kehoe concluded that his study "fails to show any evidence for the existence of such hazards."[49]

Today, looking back at the controversy of the 1920s, we may be tempted to look askance at public health professionals of the period who put their faith in the ability of scientific investigations to settle the issue. After all, those like Alice Hamilton and Yandell Henderson who fought the introduction of lead into gasoline were the strongest advocates of an impartial, governmentally-sponsored scientific study to determine the safety or dangers of tetraethyl lead. In addition to this study, the opponents of tetraethyl lead also won a ban that was to extend until the scientists' results were made available. What went wrong? Why is tetraethyl lead still a prime source of lead in the environment? Of course, there were those who had such an ideological commitment to industrial progress that they were willing to put their science aside to meet the demands of corporate greed. But more importantly, we should look at those who considered themselves honorable scientific investigators; ultimately, they could not distinguish between their "science" and the demands of an economy and society that were being built around the automobile. Any boundaries between science and society, if they ever really existed, broke down when they agreed to conduct a short-term study that would provide quick answers guaranteed, in retrospect, not to disrupt this vital industry. The symptoms of lead accumulation due to exhaust emissions would be unlike anything they had previously encountered in industrial populations, but because of compromises in their experimental design they could not possibly understand what we now know: that those most affected would not be adults, but children slowly accumulating lead. Their suffering is all the more tragic because of the amorphous and still poorly understood effects of lead on the nervous systems of children. The best of the public health scientists of the 1920s were working from an inadequate model of disease causation, but their inability to draw conclusions valid by modern standards speaks more to the interlocking relationships between science and society than to the absence of a link between lead and disease.

N O T E S

1. G. W. Mielke et al., "Lead Concentrations in Inner-City Soils as a Factor in the Child Lead Problem," *American Journal of Public Health* (AJPH) 73 (December 1983): 1366–69; Kathryn R. Mahaffey, editorial, "Sources of Lead in the Urban Environment," *AJPH* 73:1357–58. The authors would like to thank William Graebner for sharing his own research and ideas with us.

2. William Mansfield Clark to A. M. Stimson, Oct. 11, 1922, NA, RG 90, USPHS, Health Service; Alfred P. Sloan, Jr., "My Years with General Motors," in *America in the Twenties*, ed. Paul Goodman and Frank Gatell (New York: Holt, Rinehart and Winston, 1972) pp. 34–50; Charles F. Kettering, *The New Necessity* (Baltimore: Williams and Wilkins, 1932) pp. 73–79; Joseph C. Robert, *Ethyl* (Charlottesville, VA: University Press of Virginia, 1983).

3. William Mansfield Clark, Memorandum to Assistant Surgeon General A. M. Stimson (through the Acting Director, Hygienic Laboratory), Oct. 11, 1922, NA, RG 90, (Public Health Service) see also N. Roberts to Surgeon General, Nov. 13, 1922,

NA, RG 443 (National Institutes of Health), General Records, 0425T Box 23 for further statements on the fears of TEL contamination.

4. A. M. Stimson to R. N. Dyer, Oct. 13, 1922, and Dyer to Surgeon General, Oct. 18, 1922, NA, RG 90.

5. Memorandum, G. W. McCoy to Surgeon General, Nov. 23, 1922, NA, RG 90.

6. H. S. Cumming to P. S. DuPont, Dec. 20, 1922, and Thomas Midgley, Jr., to Cumming, Dec. 30, 1922, NA, RG 90.

7. A. C. Fieldner to Dr. Bain, Sept. 24, 1923, NA, RG 70 (Bureau of Mines), 101869, File 725.

8. S. C. Lind, Chief Chemist, to Superintendent Fieldner, Pittsburgh, Nov. 3, 1923, and Fieldner to Lind, Nov. 5, 1923, NA, RG 70, 101869, File 725.

9. A. C. Fieldner to Dr. Bain, Sept. 24, 1923, NA, RG 70, 101869, File 725.

10. "Agreement between the Department of Interior and General Motors Chemical Company, Dayton, Ohio," NA, RG 70, 101869, File 725.

11. C. A. Straw to Sayers, Aug. 22, 1924, NA, RG 70, 101869, File 725.

12. Graham Edgar to Dr. Paul Nicholas Leech, July 18, 1924, NA, RG 70, 101869, File 725.

13. Yandell Henderson to R. R. Sayers, Sept. 27, 1924, NA, RG 70, 101869, File 725.

14. C. W. Deppe to Hubert Work, Oct. 31, 1924, NA, RG 70, 101869, File 775.

15. New York Times, Oct. 27, 1924, p. 1; see also New York World, Oct. 27, 1924, pp. 1, 6: "When M. D. Mann, head of the research department . . . was asked . . . he denied the men had been affected by the gas."

16. New York Times, Oct. 28, 1924, p. 1.

17. New York Times, Oct. 29, 1924, p. 23; New York World, Oct. 29, 1924, p. 1.

18. New York Times, Oct. 30, 1924, p. 1.

19. New York Times, Oct. 31, 1924, p. 1.

20. New York State Department of Health, "Health News," Feb. 2, 1925, NA, RG 90, General Files, 1924–1935, 1340–216, Tetraethyl Lead.

21. New York Times, Nov. 1, 1924, p. 1.

22. E. E. Free to R. R. Sayers, Oct. 21, 1924, NA, RG 70, 5445, File 437; Reports of Investigations—Department of the Interior—Bureau of Mines; R. R. Sayers et al., "Exhaust Gases from Engines Using Ethyl Gasoline," NA, RG 443, General Records, 0425T.

23. C. K. Drinker to Sayers, Jan. 12, 1925, Sayers to Drinker, Jan. 15, 1925, and Drinker to Sayers, Jan. 19, 1925, NA, RG 70, 101869, File 725.

24. Hamilton to Surgeon General Cumming, Feb. 12, 1925, NA, RG 90, General Files, 1924–35, 1340-216, Tetraethyl Lead.

25. Workers' Health Bureau to Emery Hayhurst, and Hayhurst to Workers' Health Bureau, Oct. 29, 1924, NA, RG 70, 101869, File 725.

26. New York Times, April 22, 1925.

27. See for example Hayhurst to Sayers, Sept. 29, 1924, NA, RG 70, 101869, File 725, in which he signs his letter "Consultant to Ethyl Gasoline Corporation."

28. Hayhurst to Sayers, Sept. 29, 1924, and Hayhurst to Yant, Oct. 4, 1924, NA, RG 70, 101869, File 725.

29. Hayhurst to Sayers, Feb. 7, 1925, Sayers to Hayhurst, Feb. 13, 1925, and Hayhurst to Sayers, Feb. 13, 1925, NA, RG 70, 101869, File 725; see also "The Workers' Health Bureau Report on the Hazards Involved in the Manufacture, Distribution and Sale of Tetra-Ethyl Lead," [April, 1925] C.-E.A. Winslow Manuscript, Yale University, Box 102, Folder 1838, 7 pp.

30. Sayers to Hayhurst, Feb. 13, 1925, and Hayhurst to Sayers, Feb. 24, 1925,

NA, RG 70, 101869, File 725; "Ethyl Gasoline," editorial, *AJPH* 15 (March 1925): 239–40.

31. "The Workers' Health Bureau Report," Winslow ms.

32. *New York Times*, June 22, 1925, p. 3. The *Times* reported that 8 workers had died in the Deepwater plant.

33. Haven Emerson to Cumming, Feb. 9, 1925, NA, RG 90, General Files, 1924–35, 1340-216, Tetraethyl Lead.

34. See Cumming, Memorandum for Dr. Stimson, Feb. 13, 1925, NA, RG 90, General Files, 1924–35, 1340-216, Tetraethyl Lead.

35. *New York World*, May 1, 1925, p. 1.

36. United States Public Health Service, "Proceedings of a Conference to Determine Whether or Not there is a Public Health Question in the Manufacture, Distribution or Use of Tetraethyl Lead Gasoline," *Public Health Bulletin #158*, (Washington, DC: GPO, 1925), p. 62.

37. Ibid., pp. 4, 69, 105–107: Howard went on to raise the question of how to resolve the problems inherent in the public health dangers of TEL. He called for the Public Health Service to decide on the future course of research and development. See also T. Midgley, "Tetraethyl Lead Poison Hazards," *Industrial and Engineering Chemistry* 17 (August 1925): 827–28.

38. *New York Times*, May 7, 1925; *New York Times*, Oct. 28, 1924; "Sober Facts About Tetra-Ethyl Lead," *Literary Digest* 83 (Dec. 27, 1924): 25–26.

39. *New York Times*, Oct. 28, 1924, p. 1; United States Public Health Service *Bulletin #158*, p. 12; *New York World*, May 9, 1925; see also *New York Times*, Nov. 27, 1924, p. 14 for statement by the American Chemical Society: "Perhaps the greatest hazard is the indifference which not only workmen but even chemists come to have for dangerous work with which they are familiar."

40. USPHS, *Bulletin #158*, p. 62; *New York Times*, April 22, 1925; Henderson to Sayers, Jan. 20, 1925, NA, RG 70, 101869, File 725, where he points out the known dangers of lead to printers, painters, and other industrial workers: "I read your paper on 'Exhaust Gases from Engines Using Ethyl Gasoline' with much interest. I note that you compare the risk to that which is faced by painters. I think this comparison is very well taken, for practically all painters suffer, at one time or another, and in some cases repeatedly, from lead poisoning." See also Grace M. Burnham, Director, Workers' Health Bureau to Editor, *New York Times*, July 6, 1925: "If the expectations of Standard Oil officials that 15,000,000,000 gallons would be sold in the next year proves true, it would mean that 50,000 tons of lead would be distributed over the streets of the country."

41. USPHS, *Bulletin #158*, pp. 60, 109; see also for the American Federation of Labor's concern: *American Federationist* 30 (August 1923): 632–33.

42. USPHS, *Bulletin #158*, pp. 108, 96; Henderson to Sayers, Jan. 20, 1925, NA, RG 70, 101869, File 725.

43. USPHS, *Bulletin #158*, p. 79.

44. Ibid., p. 98; Alice Hamilton, "What Price Safety, Tetraethyl Lead Reveals a Flaw in our Defences," *The Survey Mid-Monthly* 54 (June 15, 1925): 333–34; *New York World*, May 22, 1925, p. 1.

45. USPHS, *Bulletin #158*, pp. 86–87, 89; "Perils and Benefits of Ethyl Gas," *Literary Digest* 85 (April 18, 1925): 17; J. H. Shrader, "Tetra-Ethyl Lead and the Public Health," *AJPH* 15 (March 1925): 213–14.

46. Flinn to Sayers, May 11, 1925, Hayhurst to Sayers, May 14, 1925, NA, RG 70, 101869, File 725.

47. Alice Hamilton, "What Price Safety," *The Survey Mid-Monthly* 54 (June 15, 1925): 333; for further elaboration of the committee members' positions see "Conference of Tetraethyl Lead," *Automotive Industries* 52 (May 7, 1925): 835; "Tetraethyl Lead Sales Are Suspended," *National Petroleum News* 17 (May 27, 1925): 37.

48. *New York Times,* Jan. 20, 1926, p. 13. See also Winslow's handwritten comments on the draft of the committee's report: he wanted a specific statement that "a more extensive study was not possible in view of the limited time allowed to the committee." Winslow MSS, Box 101, Folder 1805, Yale University. See also other committee members' responses to the draft, Folders 1800 and 1801. For further elaboration of the Committee members' positions see: "Conference of Tetraethyl Lead Gasoline Committee—Afternoon Session," Dec. 22, 1925, NA, RG 90, General Files, 1924–1935, Box 109. We are grateful to William Graebner for pointing out this document to us. For a copy of this report itself, see: Treasury Department, USPHS, "The Use of Tetraethyl Lead Gasoline in its Relation to Public Health," *Public Health Bulletin #163* (Washington: GPO, 1926).

49. Kehoe et al., "A Study of the Health Hazards Associated with the Distribution and Use of Ethyl Gasoline," April, 1928, from the Eichberg Laboratory of Physiology, University of Cincinnati, Cincinnati, OH, NA, RG 70, 101869, File 725.

Hegemony through Science: Information Engineering and Lead Toxicology, 1925–1965

For almost a half century, the voluntary rules of 1926—the codified legacy of the Bayway, New Jersey, tragedy—remained the only regulatory framework governing leaded gasoline. Not until the late 1950s, when the fuel industry began to consider adding more lead to gasoline to satisfy the more powerful engines that were on the drawing boards, did the subject again reach the level of public discussion; and it would be another fifteen years—1973—before the new Environmental Protection Agency would issue its first regulations on lead emissions.

How does one explain this long period of acquiescence? One possibility is that there was no need for further regulatory activity; that the regulations of 1926 had eliminated hazardous conditions in the manufacture and distribution of leaded gasoline. There is undoubtedly a certain validity to this argument. Though ugly stories continued to circulate about the hazards of working with tetraethyl lead (one, out of Charleston, South Carolina, in the early 1930s, had men "losing their arms and legs" through contact with the additive), there seems to have been no recurrence of the Bayway disaster until 1960.[1] This is not to say that all occupational hazards associated with the production and distribution of tetraethyl lead had been eliminated; only that workers did not die in bunches.

A second explanation for the period of inactivity also has to do with the way in which the crisis of the 1920s was handled and resolved. As effective as the 1926 regulations were, they were occupational rather than environmental controls, unrelated to automobile emissions. Although Yandell Henderson, the *New York World*, and others had attempted to shift the focus of public debate from the New Jersey factories to the streets of New York City—that is, from occupation to environment—their efforts were for naught. Robert Kehoe's occupational perspective had triumphed.

If lead had proven to be of little import environmentally, then the period of inactivity would be most easily explained: viz., no activity occurred because none was necessary. In fact, the opposite happened: lead came to be understood as being of great environmental significance. An early breakthrough toward an environmental perspective took place in 1955, when a study of Philadelphia tenements revealed that the city's children were becoming ill and dying from eating chips of lead-based paint. While the paint-chip problem was not environmental in the classic sense—there was no contamination of air, water, or the *normal* food supply—it was also obvious that this situation did not involve occupation. Another sign of the progress of environmentalism could be seen during the 1960s, when jokes about Los Angeles smog—jokes that implied that Los Angeles had a special, even humorous problem that the rest of the nation had somehow avoided— were replaced by a growing recognition that automobile emissions were a real or potential hazard in every population center. In the 1970s and 1980s, scientists delivered the most damaging blow to the old occupational perspective; studies of children established a clear link between the use of leaded fuel and high levels of lead in the blood, and another link between elevated lead blood levels and problems in thinking and learning.[2]

Why had it taken so long, then, to confirm that environmental lead was a legitimate hazard? Put somewhat differently, why did those institutions, agencies, and individuals charged with studying such matters take so long to find the truth?

The single most important answer to that question is that the lead industries did not want to see the triumph of an environmental perspective; and the lead industries exercised enormous influence over the production and dissemination of knowledge about lead in the four decades after 1925. This influence might best be described as a kind of hegemony over scientific research and over perceptions of lead-related problems. Hegemony was achieved through a variety of institutions, including the Lead Industries Association, a trade association; the Mellon Institute, a private research organization; the Charles F. Kettering Foundation and Kettering Laboratory of Applied Physiology; and the American Public Health Association. Through these institutions, the lead industry and its allies in autos and oil underwrote and carried out much of the significant lead-related scientific research in the inter-war years and beyond. Scientists whose work was thereby supported were, in turn, influential in determining how Americans thought about and responded to lead hazards. In addition, the lead industries were able to achieve and maintain this hegemonic control because no challenge was mounted to their point of view. Public agencies and professionals, organized in the American Medical Association, the American Association of Industrial Physicians and Surgeons, the American Public Health Association, and the United States Public Health Service, failed to provide either an alternative perspective or a foundation of reasonably objective scientific knowledge about lead toxicology on which governmental or private actions might be based.

ROBERT KEHOE, CHARLES KETTERING, AND THE KETTERING FOUNDATION

The most significant links between the business community and the scientific establishment involved the Charles F. Kettering Foundation, General Motors, and Robert Kehoe. It would be difficult to overestimate the importance of these ties, because Kehoe was until the mid-1960s the nation's most vocal and influential scientist working on lead hazards. The Surgeon General's 1925 conference on tetraethyl lead was only the first of Kehoe's numerous public appearances. As late as 1966, Senator Edmund Muskie could be seen grappling with Kehoe's science and its implications during hearings on air pollution.[3] Kehoe was by then nearing the end of his reign, but he had ruled the scientific domain long enough to significantly shape the nation's response to the hazard of lead.

Kehoe was the first director of the Kettering Laboratory of Applied Physiology, constructed with Kettering Foundation funds at the University of Cincinnati and opened in 1930. Trained as a physician, Kehoe had worked as a hospital pathologist in the early 1920s while holding down his first teaching job as Assistant Professor of Physiology at the University of Cincinnati. In 1925 he added the post of Medical Director of the Ethyl Gasoline Corporation (whose executive officers included Kettering and Thomas J. Midgley, Jr.). Kehoe simultaneously held official positions with the laboratory, the University of Cincinnati, and the Ethyl Corporation until 1958, when his relationship with the Ethyl Corporation ended. Frank Princi, Kehoe's successor as Director of the Kettering Laboratory, also enjoyed this triple appointment.[4]

Kehoe's education in lead toxicology began in early 1925, when the Ethyl Corporation's new Medical Director investigated 138 cases of poisoning—attributable to tetraethyl lead—in Dayton, Ohio and Bayway. A year later, Kehoe helped draft the regulations governing the blending of gasoline in refineries, and he was thereafter responsible for the supervision of those regulations.[5]

No person was more important in the bonding that occurred between corporate and scientific communities than Charles Kettering. At the age of thirty-three he had left the National Cash Register Company for a position of partial ownership in the Dayton Engineering Laboratories Company, known as Delco. In 1916, Kettering sold his interest in Delco to a company that later became part of General Motors, remaining as an operating head. When GM felt the need for its own research laboratory, Kettering agreed to head it, provided the facility would be located in Dayton. After 1920, Kettering was at the center of GM's scientific and technological efforts, including Midgley's experiments with tetraethyl lead as a gasoline additive.[6]

Kettering was no less adept at making money than at invention, and in 1927 his savings found an outlet in the Kettering Foundation. Created to "sponsor and carry out scientific research for the benefit of humanity," the foundation over the years funded programs in photosynthesis, biology, energy conservation, nutrition, and, beginning in 1940, cancer research. The General Motors connection was enhanced in 1945, when Alfred P. Sloan, Jr., GM's first Chairman of the Board, provided the funds for what became the Sloan-Kettering Institute for Cancer Research.[7]

By 1933, a substantial body of work bearing Kehoe's name or imprint had appeared in the industrial hygiene and medical journals. In these early articles, Kehoe laid out the Kettering-Kehoe perspective on lead poisoning. The following arguments were central to that perspective. First, small amounts of lead occurred "naturally" in human tissue, even in persons living in rural areas entirely divorced from modern industrial processes and automobiles (this claim was based on a study of two communities in rural Mexico). Therefore, to discover the presence of lead in human excreta or tissue was not proof that the subject had encountered some alien source of lead, but only that he had lived life "on a lead-bearing planet."[8]

Second, this natural accumulation of lead in the body, caused largely by the ingestion of vegetation, did not result in the storage of greater and greater amounts of lead in the tissues. Instead, an "equilibrium" was established between lead intake and lead elimination. Beyond that point of equilibrium, absorption did not occur. The body, it seemed, had a built-in protective mechanism. Third, because lead occurred naturally in human tissue even in primitive societies, and because its storage occurred within an equilibrium, there was no necessary relationship between lead absorption and lead intoxication—no necessary correlation between lead concentration in the feces, urine, or tissues, and lead poisoning.[9]

For Kehoe, clinical findings were insufficient; one needed to know "that a significant exposure to lead has occurred."[10] In practice, this led directly to the fourth point. Because general, environmental exposures to lead could be measured only clinically, and because clinical findings were inadequate, hygienists should eschew environmental hazards—such as those potentially attributable to lead emissions—in order to focus on occupational lead poisoning. Although Kehoe acknowledged the possible long-term health consequences of an increased daily intake of lead (while doing his best to minimize the relationship between lead poisoning and intake), this was clearly not the thrust of his work.[11]

Thus Kehoe, the Kettering Laboratory and Foundation, General Motors, and the Ethyl Corporation had joined hands to create a body of scientific opinion that was ideally suited to the manufacturers of both tetraethyl lead and automobiles. Kehoe's work said that industrialization (viz., the automobile, leaded gasoline) was not responsible for lead in the body; that even if it were, it was not harmful; and that even if it were harmful, a cause-effect relationship was impossible to establish. Kehoe's science also undermined modern laboratory testing and made it seem as if a sure diagnosis of lead poisoning could be made only if one matched extreme symptoms with an occupation recognized as highly dangerous.

Of course, even these arguments obligated the Ethyl Corporation to clean its own house, and it had done so with the 1926 regulations. But this was a small price to pay for insurance against the only real threat—the threat that lead poisoning might be environmentally defined and that tetraethyl lead might then be banned.

Kehoe pressed his science into every nook and cranny of industrial hygiene, public health, medicine, and science. He served on the 1930 American Public Health Association Committee on Lead Poisoning, whose report lashed out at legislation and at "arbitrary rules by government agencies" and focused on occupational problems and on diagnosis as a virtual art form. When the APHA prepared a report on how to determine the lead content in air and biological materials, another Kettering employee, Jacob Cholak, did the work. Kehoe was a member of the American Standards Association committee that developed the 1943 lead standard, and he was available to testify for companies involved in lead compensation cases. He had ties to the American Association of Industrial Physicians and Surgeons, the American Medical Association, the American Industrial Hygiene Association (he was a Director from 1940–43), and the National Research Council.[12] Kehoe and Cholak were consultants on the 1952 *Air Pollution Abatement Manual* of the Manufacturing Chemists' Association, a document that described air as a "natural means of disposing of useless residues" and remarked that scientific proof of a relationship between air contaminants and damage to health had been "elusive."[13]

Kehoe was still hard at work in the mid-1960s, when airborne lead again became an issue, this time under the rubric of air pollution. He could be found at a 1965 USPHS symposium on lead contamination, summarizing and defending a life's work.[14] A year later he was filling Muskie's ear in a Senate hearing room, brandishing his 1933 studies as if they were newly forged swords. "The situation is in no sense urgent," he replied when asked about lead contamination of the atmosphere, and when Muskie asked him to comment on the desirability of a substitute for lead in gasoline, Kehoe managed to fishtail his way out of the question. Kehoe's repertoire had been fortified with a new, 30-year study of Cincinnati's air and lead pollution which purportedly demonstrated a slight decrease in airborne lead levels. Kehoe had coupled this new data with his forty-year-old emphasis on occupations. One might, he argued, be concerned about persons who drove trucks or delivery wagons, and who therefore spent considerable time in the proximity of lead emissions. "The point with respect to human exposure," he emphasized, "is that there are really not very many people who spend much of their lifetime on a freeway."[15] Kehoe was vigilant in defense of this occupational perspective.

The dominance of the Kettering/Kehoe perspective was a well-known fact within American medical science. At the 1965 Public Health Service symposium, a Harvard University physiologist singled out the Kettering Laboratory. "It is extremely unusual in medical research," he said, "that there is only one small group and one place in the country where research in a specific area of knowledge is exclusively done." The following year the Public Health Service finally acknowledged the heavy industry sponsorship of lead research.[16]

Kehoe did not deny it. He freely admitted that most of the research on lead and the environment had been funded by the manufacturers and distributors of tetraethyl lead, and that much of it had been carried out in the Kettering Laboratory and in the Department of Preventive Medicine and Industrial Health of the College of Medicine at the University of Cincinnati.[17]

THE MELLON INSTITUTE

The Kettering Laboratory was unique in the breadth and depth of its influence. But there were other, similar institutions that also shaped the scientific product. The Mellon Institute was founded in 1913 with funds from Andrew W. Mellon and Richard B. Mellon. The Mellons covered the Institute's operating deficits until 1921, when the Pittsburgh-based facility became "self-supporting" through industry sponsorship. Its first interest in lead surfaced in 1925, when a Mellon Institute representative (who was also a Gulf Refining Company employee) appeared at the Surgeon General's conference on tetraethyl lead. Two years later, the Institute was incorporated for "long-range research in pure and applied science."[18]

The Institute sponsored field and laboratory research on its premises and also functioned as an umbrella for organizations directly involved in the politics of industrial and environmental health. In the 1950s, for example, the Institute provided office space and other facilities free of charge to the Air Pollution Control Association (APCA).[19]

The APCA dates to the Progressive Era, when it was a smoke control association with a membership of city smoke inspectors. The modern version of the association may be traced to 1936, when R. R. Sayers (responsible for the controversial early-1920s Bureau of Mines research on lead) presented a paper, "Atmospheric Pollution in American Cities," at the annual meeting in Atlanta. Although city officials and smoke abatement types remained influential within the organization for many years, by the mid-1950s, when the Mellon Institute offered its support, the APCA was catching up with the rapidly changing science of air pollution, and its membership had come to include some 175 corporations with an interest in pollution. In the early 1960s, the Association's Board of Directors included technical people from Gulf Oil in Pittsburgh and National Steel Corporation in Weirton, West Virginia—a further indication that the APCA was in transition from the urban regulatory, public-health emphasis of the Progressive Era to a pro-business stance on industrial air pollution.[20]

Unlike the Air Pollution Control Association, the Industrial Hygiene Foundation (IHF) was from the beginning linked to the Mellon Institute. Originally called the Air Hygiene Foundation, it was created in 1935 in connection with the Mellon Institute and around the issue of silicosis. Some twenty members included Allegheny Ludlum Steel, the Aluminum Company of America, Du Pont, Owens-Illinois Glass, the Public Health Service, and the Bureau of Mines. The name was changed to Industrial Hygiene Foundation in 1941. Like the APCA, this organization was headquartered at the Mellon Institute.[21]

The IHF was described in 1947 as a national nonprofit research organization for the "advancement of employee health in industry and the improvement of working conditions."[22] But this description avoids the organization's integration of business and science. The money for IHF programs came from business—more than 335 corporate sponsors in 1947—and the organization's Board of Trustees, dominated by corporate officers from American Brake Shoe Company, U.S. Smelting, Refining & Mining Company, and Employers Mutual Liability Insurance Company, reflected these contributions.[23]

Naturally, the Foundation's informational and research programs did not offend its business sponsors. Thus the vice president of the Aluminum Company of America could wax eloquent over the Foundation's contributions to free enterprise. "Industry," wrote John D. Harper, "can feel justly proud of the Industrial Hygiene Foundation. It is encouraging to me that in an era when the expansive grasp of government is reaching out to regulate and control more and more phases of economic life, there is a voluntary and nongovernmental research organization to turn to on problems of environmental engineering." Andrew Fletcher, board chairman of the IHF, offered less doctrinaire praise for the Foundation's efforts at the November 1955 annual meeting, applauding the organization for its role in stimulating and encouraging health maintenance in industry. Fletcher was president of the St. Joseph Lead Company.[24]

The IHF's credibility as a developer and purveyor of scientific and technical knowledge depended on the appearance, at least, of balance and objectivity. This appearance proved not at all difficult to achieve, for the scientific community included many who agreed philosophically with the conservative ideology of the foundation or who were willing to lend their names to its endeavors. Although the Board of Trustees was dominated by business, it also included Philip Drinker of the Harvard School of Public Health and Dan Harrington of the Bureau of Mines.[25]

In the late 1950s and early 1960s, the IHF was as dedicated as ever to a business viewpoint; yet the organization was run by a former government scientist of some reputation, H. H. Schrenk. A University of Wisconsin Ph.D., Schrenk served as Chief Chemist of the Health Division of the Bureau of Mines from 1936 to 1945, then briefly as Chief of the Environmental Division of the Public Health Service. In 1949 he joined the IHF as Research Director, and took over the post of Managing Director in 1958. During his career as a government scientist, Schrenk had served as president of the American Industrial Hygiene Association and as chairman of the Pittsburgh Section of the American Chemical Society, and he had published more than 100 professional papers on a variety of subjects, including lead poisoning.[26]

Schrenk was at once scientist, manager, and advocate of free enterprise capitalism. He had come to the IHF because it too integrated these roles. In 1963, he described his impatience with those who "exaggerate and dramatize accidental occurrences and alleged injurious efforts which have not been established." These "attacks" on reasonable industrial practices could be met, Schrenk believed, with "expanded research and education." Through its information services, toxicologic research, field investigations, and seminars, the Industrial Hygiene Foundation would contribute to "the positive side of the ledger."[27]

What was the "positive side of the ledger" with relationship to lead poisoning? First, it meant acceptance and development of Kehoe's work on lead absorption. Schrenk and another IHF scientist agreed that lead was everywhere, and that it existed in the human body without any necessary adverse effects. Second, it meant suspicion of "standards"—i.e., standards for lead absorption in the body and environmental standards for air quality. Standards served a purpose, but they had to be measured against "the response or lack of response of the worker," and that could only be accomplished by thorough medical surveillance. "Figures," Schrenk concluded, "are not a substitute for sound judgment." This shifted responsibility for the determination of lead poisoning from a "standard" established by an outside agency to the medical officer in the corporation.[28] It was an argument the IHF's corporate sponsors must have found congenial.

THE LEAD INDUSTRIES ASSOCIATION

On the surface, the Lead Industries Association (LIA) was a very different sort of organization from the Industrial Hygiene Foundation. It was a trade association, pure and simple, unabashedly devoted to telling the story of lead from the perspective of the corporations that produced and used it. As a result, the LIA functioned as a public relations office. Its journal, *Lead,* ran one article after another extolling the virtues of the element. One celebrated the 130 miles of lead pipe that carried New Bedford's water supply; another praised the lead paint that protected a bridge over the Hudson River; a third triumphantly announced, "Use of Tetraethyl Lead Growing Rapidly."[29] For a time in the late 1930s, the Association ran a series of ads on the back of the journal in which those most likely to be victims of lead poisoning could be found lauding lead's virtues. One of these featured a lead miner, light on his cap, talking about lead paint. "I'm a miner not a painter," he said. "The metal I mine out of the earth is lead. And mister that lead is what gives life and gumption to paint. You think I'm prejudiced? Ask any person who's been at it long enough to see how his work stands weather. He'll tell you the same. . . . So some of the real good painters are boosters for white lead paint. They know that the way a white lead job stands up helps to build their reputation."[30]

On closer examination, however, the aims and methods of the LIA appear remarkably like those of the IHF. Like the IHF, the LIA realized that the most effective defense of lead was one underpinned by science and presented as science. The Lead Institute, apparently a predecessor organization, had embarked on this course in 1925, when it gave funds to the Harvard Medical School in support of Joseph C. Aub's study of lead.[31] Although the grant was ostensibly made without strings, Aub's work proved supportive of the industry in several ways. First, while Aub did not argue that lead was present in all human tissues, he did emphasize the importance of determining lead absorption levels for people "living a usual normal life at a given time and place." This implied that a certain level of lead absorption might be "normal," if not "natural." Second, Aub provided a primitive statement of a notion later more fully developed by Kehoe, that the presence of lead in the excreta was less a sign of *poisoning* than evidence of a healthy *elimination* of lead. Third, he observed that much of the lead absorbed by the body was not eliminated but stored in the bones, and was, in that state, "apparently harmless. . . ."[32] Decades later, Aub's work was being used to support the need for company physicians to distinguish clearly between simple lead absorption and lead intoxication, or "true lead poisoning."[33]

After the LIA was established in 1930, the Association helped finance the Kettering abstracting service and research at Children's Medical Center in Boston, the Harvard Medical School, and the University of Cincinnati.[34] Some of the Cincinnati money surely found its way to the Kettering Laboratory and to Kehoe, who from the beginning was something of an industry hero. Year after year, at LIA meetings, at national conferences on lead poisoning, and in testimony before Congressional meetings, Kehoe's name was invoked with religious regularity. Kehoe's findings and their implications, nearly always supportive of the lead industry, could be heard at every turn: Kehoe had proved the existence of a lead "equilibrium"; he had shown that people could handle small amounts of lead without ill effects; he had demonstrated the need to avoid knee-jerk diagnoses of lead poisoning based solely on laboratory data.[35]

But the LIA's work did not depend on Kehoe, the Kettering Laboratory, or Aub's book. There was plenty of research being done that buttressed the industry's position; the industry had only to ensure that it received a public hearing and took on the aura of science. This the LIA accomplished through an ongoing series of lead conferences. The stated function of these conferences was to "bring together the qualified minds of the country most likely to reach a consensus as to optimal measures in the prevention, the detection and the therapy of lead poisoning."[36] Invariably, however, LIA gatherings assembled not the most qualified but those whose work most clearly supported the lead industries. The University of Cincinnati/Kettering Laboratory was always well represented. Another regular speaker was Elston L. Belknap, Associate Professor of Medicine at Marquette University and a consultant to the Globe-Union Company. Belknap could be found, like Kehoe and others, pointing out the dangers of diagnosing lead poisoning purely from laboratory tests. And Philip Drinker, a chemical engineering Ph.D. teaching at Harvard's School of Public Health, was only preaching to the converted when he told a 1946 gathering that the dangers of lead paint, lead-painted toys, leaded gasoline, and lead plumbing had virtually ceased to exist, and that "lead poisoning among the population at large is rare. . . ."[37]

More was involved here than a trade association defending its corporate members. That defense was accomplished through a format—the conference—that was scholarly, objective, and "scientific" in appearance. In addition, when scholars such as Drinker and Kehoe presented their research in such corporate forums, they allowed the lead industries to associate themselves all the more fully with the world of science and thus to further justify their point of view.

As was the case with the Industrial Hygiene Foundation, the leadership of the LIA perfectly represented this fusion of industry and science. For many years the Association was headed by Felix E. Wormser. An executive with the St. Joseph Lead Company, Wormser had helped organize the LIA in the late 1920s, when public fear of lead poisoning ran high. His approach was pure public relations: replace the bad news with the good. In 1946, reacting to a story that had appeared in the *New York Daily News* with the photo caption "Lead Fume Victims," Wormser described his philosophy of public relations. "Unfortunately," he said, "this is not the type of publicity relished by the lead mining and fabricating industries. They do not mind reading items describing the indispensable tasks performed by lead from day to day in making civilized life comfortable, but they do object to the unfavorable publicity that lead receives occasionally when it is absolutely innocent of any wrongdoing."[38]

Yet Wormser was not just a businessman. He held a degree in mining engineering, was a fellow of the American Association for the Advancement of Science, and served as Assistant Secretary of the Interior for Mineral Resources under Eisenhower. He also understood that to be effective, lead's defense would have to be based on something more substantial than the unvarnished opinion of the industry that lead was harmless. The position of the Association must *appear* to be—or *actually* be, depending on how one conceptualizes Wormser's ideology—derived from knowledge, science, and truth itself.[39]

To ensure the viability of this link between the lead industry and science, Wormser's association subvented particular research projects and sponsored scientific conferences. Wormser followed the medical and scientific community closely. He knew which committees were likely to be sympathetic to the lead industry's position, and he knew which organizations might be counted on to produce pro-industry reports. He cultivated relationships with the

University of Cincinnati, the Public Health Service, the American Public Health Association, the United Nations World Health Organization, and the American Medical Association, through its Council on Industrial Health.[40]

This familiarity allowed Wormser to use public organizations to serve the needs of the lead industry. In 1948, for example, Wormser described a crisis in the apple industry brought on by the use of a lead insecticide. Wormser had been informed that unless a reasonable tolerance for lead was set, the industry could not function. "I urged a complete and impartial investigation," Wormser recalled, "and, because of the public interest involved, suggested the use of the U.S. Public Health Service . . . this was subsequently done and . . . the Public Health Service issued an excellent report. . . ."[41]

The conventional scholarly wisdom is that by the 1930s, the funding of public health was shifting from the private foundations to the federal government.[42] While this was undoubtedly true in the aggregate, the experience of the Mellon Institute, the Industrial Hygiene Foundation, and the Kettering Foundation suggests that foundations and private corporate wealth maintained a strong position in lead research and policy into the 1960s. Moreover, one should not distinguish between the foundations and other forms of private support for scientific research and its dissemination. Neither the LIA nor the Mellon Institute was a foundation—and neither, its name notwithstanding, was the Industrial Hygiene Foundation. Yet each understood the relationship between science and the business community as well as the foundations did. Together, these organizations helped the lead industry maintain a formidable presence in the science of public health.

MEDICAL SCIENCE AND PUBLIC HEALTH

At bottom, the failure of the "system" of knowledge occurred less because the corporate sector did what was natural and inevitable than because scientific and medical professionals who could reasonably have been expected to provide an alternative perspective did not do so. They were, in fact, part of the problem. Physicians and public health professionals, organized in the American Medical Association, the American Association of Industrial Physicians and Surgeons, and the American Public Health Association and lodged governmentally in the Public Health Service, did not provide the alternative perspective that lead toxicology required in the 1930s and 1940s.

In the Progressive Era, the disinterest of medical professionals in industrial hygiene and toxicology, while nearly unanimous, was certainly understandable. It was almost impossible to be formally educated in the subject. Although the first course in industrial hygiene was offered in 1905 at MIT, for decades the field was not a serious part of the curriculum at most medical schools. Not until 1918 did a university offer a degree in industrial hygiene.[43]

Moreover, few practitioners of industrial medicine—especially in the new field of toxicology—could be found outside the corporate sphere, and employment within the corporation was not conducive to social reformism, independence of mind, or even sound medical practice. The National Lead Company of Brooklyn, for example, employed one Lawrence Paipken as physician in its Bradley and Atlantic Works. Paipken held this position in addition to his private practice, so he spent only two hours a day at the plant, where he looked after some 500 workers. Paipken's knowledge was in every respect inadequate. He saw only those cases of lead poisoning sent to him by company foremen, and he knew almost nothing of the work done in the facility or of the health and safety technology, primitive as it was, in use at the plant. From such evidence as this, Alice Hamilton concluded that for a physician to take up a relationship with a private business was to "earn the contempt of his colleagues."[44]

The unsavory reputation of the occupation of company physician and the growing prominence of industrial health and safety were factors in the professionalization of industrial medicine. The American Association of Industrial Physicians and Surgeons (AAIPS) was founded in 1916; almost all of its members were from industry, though a few were part of a public health "faction." Several hundred corporate medical departments formed the backbone of the association.[45] From the beginning, the AAIPS, and indeed the entire discipline of industrial medicine, saw itself in the dual role of advocating efficiency engineering on the one hand, and some brand of medical science on the other. Industrial physicians were, in other words, scientific managers with medical degrees.[46]

In 1938 the AAIPS attempted to reach out to non-physicians and create an umbrella organization for all those interested in industrial hygiene. The new association, called the American Industrial Hygiene Association (AIHA), was assembled through the efforts of Carey McCord. Since 1920 McCord had been Director of the Industrial Health Conservancy Laboratories, a private Cincinnati organization that served as consultant to corporations and government agencies. He had chaired the APHA Industrial Hygiene Section and had helped write that group's 1930 report on lead poisoning. In the 1940s he served as a Director of the American Industrial Hygiene Association, and in the 1950s he would author an important history of lead poisoning.[47]

As finally constituted, the AIHA was remarkably inclusive. Among its approximately 175 members in 1939 were physicians from large corporations (the heart of the old organization of industrial physicians) and representatives from schools of medicine, engineering departments in corporations, insurance companies, private research laboratories (including Kettering), state departments of labor, the U.S. Bureau of Mines, and Harvard's School of Public Health. Unlike the parent body of physicians, the new organization's broader constituency promised to give it credibility and influence beyond that of its predecessor.[48]

Yet the formation of the new organization did not portend a softening of the physicians' pro-business orientation. In fact, no sooner had the AIHA come into existence than it answered the call to cooperate with the National Institute of Health in fashioning and applying health programs that would contribute to the national defense by reducing absenteeism in industrial facilities. While better than no industrial hygiene at all, and understandable in the context of a war whose demands knew no bounds, this work put AIHA medical professionals in the role of industrial relations experts, utilizing health as a means rather than pursuing it as an end in itself.[49]

Because only those with medical degrees could belong to the AAIPS, those lacking the degree but with an interest in industrial hygiene had to turn to other organizations. Some found a place in the scientific societies; some chemists, for example, pursued lead toxicology through the American Chemical Society. Others became involved with the American Public Health Association (APHA), founded, in 1872. The APHA created an Industrial Hygiene Section and a Committee on Lead Poisoning in 1930.[50]

It is doubtful whether the APHA offered members a perspective distinct from that of the AAIPS. Although the Committee on Lead Poisoning included professors, state government officials, and others not eligible for membership in the AAIPS, its ten members in 1930 included the Medical Director of the DuPont Company and two future directors of the AAIPS, including the ubiquitous Kehoe. The reports of the Committee on Lead Poisoning were, moreover, invariably conservative. Its 1930 report, for example, found "lead poisoning" a "rather unfortunate term" that conveyed "ideas of extreme distress and dramatic episodes." This report also attacked the "arbitrary rules [laid down by] governmental agencies" and emphasized the commitment of industry, especially the larger companies, to "the economic value of humanitarianism."[51] When the APHA called on scientists to study methods for measuring lead content in air, it turned to Cholak and Kehoe of the Kettering Laboratory.[52]

The AMA's position on industrial health was self-serving. The Association's Council on Industrial Health seems to have been preoccupied with making sure that physicians got their share of the industrial health pie, a goal that produced an odd mixture of scientific and public relations functions. Speaking to the Council in 1946, its chairman, Stanley J. Seeger, revealed his fears that industrial health was slipping from the hands of the physicians. A recent agreement between the federal government and the United Mine Workers was especially disturbing, since it placed health and medical services "under union control." Assuring business that the AMA understood its needs and had "much to contribute to improved human relations in industry," Seeger emphasized the importance of placing the physician at the center of corporate health programs. He concluded by implying that the AMA could guarantee employers the right kind of scientific information.[53] Medical science on call.

The AMA's relationship to lead poisoning shows the same biases. A published symposium on lead poisoning, appearing in the *Journal* of the AMA in 1934 and 1935, reveals a virtually monolithic point of view. The Kettering Laboratory was well represented in a paper by Kehoe and in another by his associates. Aub and Roy Jones of the Public Health Service critiqued the idea of a "lead standard," Jones concluding that "all observations should be considered in their relation to the entire clinical picture." Because only physicians understood the "clinical picture," that conclusion, valid or not, was self-serving. So was that of another physician, Irving Gray, who praised the "careful" Kettering lead absorption studies and aligned himself with the Kettering position that "the diagnosis of lead intoxication must continue to rest largely upon the skill and judgment in the elicitation and interpretation of clinical evidence."[54]

The most interesting piece in the AMA symposium was by Milwaukee physician Belknap. He too insisted that lead poisoning was not "reducible" to a formula that a plant superintendent or a lay personnel officer could manipulate. More important, however, was Belknap's conclusion that industrial physicians had a special responsibility to deal with lead poisoning while keeping workers on the job. A working system for the control of lead poisoning was not "so simple as discharging a worker to swell the ranks of disabled veterans and county institution inmates." It surely involved more than "hospitalizing a man for weeks of exhaustive research" and costly compensation payments. "The actual title of this paper," Belknap concluded, "should therefore be The Control of Lead Poisoning in the Worker Actually at Work." This would be accomplished ideally by moving the leaded worker to another job, but if that were not possible, a worker could be kept in the same position and gradually deleaded by increasing his calcium intake. "The physician's work in industry," Belknap summarized, "is to deal with the practical necessity of keeping the worker well while he continues his hazardous occupation. In this way they can best cooperate with the management for every one's benefit."[55]

Nor was the United States Public Health Service (USPHS) ideally suited to serve as an unbiased and objective agency or as the voice of workers in the lead trades or industries. Early studies of the lead trades by the USPHS were not uncritical of industry, but those same studies found the Service reluctant to step even figuratively into the factories as engineers, and more reluctant still to suggest legislative remedies. The free enterprise bias of the agency was apparent from the beginning. Throughout its history, the USPHS has engaged in the questionable practice of conducting lead research when a particular lead industry or trade association wanted it to. The Service's 1937–38 study of the lead battery industry, for example, was carried out at the request of the National Battery Manufacturers' Association. A 1950 study of water-carrying pipes painted internally with lead was done for the Lead Industries Association. And an inquiry into the health effects of lead arsenic insecticides was initiated by apple growers.[56]

Although in theory all these studies could have confirmed the worst fears of those who requested them, none of them did. In a conclusion not universally accepted, the Service virtually dismissed the argument that painted standpipes could affect the public health. Apples sprayed with a lead compound were similarly exonerated, and one official of the Public Health Service could be heard commenting on the "long and honorable history" of insecticides containing lead in modern agriculture.[57] The battery industry study uncovered "no cases of plumbism severe enough to cause disability" and, like the AMA, came down against the diagnostic significance of blood and urinary lead concentrations. Another controversial study, of atmospheric lead, was managed by the USPHS but actually conducted in cooperation with the automobile industry, fuel producers, and the Kettering Laboratory. The study concluded that 1961–62 levels of airborne lead were below those of a quarter century earlier.[58]

In addition, one could question the independence of some agency personnel. Royd R. Sayers, who in the 1940s became medical director of the Public Health Service, had authored the controversial Bureau of Mines study on tetraethyl lead in the early 1920s, and he had served as president of the business-dominated AAIPS in 1937–38. The USPHS's Environmental Investigations Branch was headed for a time in the 1940s by H. H. Schrenk, who in the 1950s would direct the Mellon-sponsored Industrial Hygiene Foundation.[59]

CONCLUSION

More than four decades after the Bayway, New Jersey tragedy caused by tetraethyl lead, geo-chemist Clair Patterson appeared before a 1966 Senate subcommittee. His testimony signified a growing appreciation within the academic scientific community of the forces that had delayed the emergence of a cautious, environmental approach to lead hazards. Patterson's special contribution to this new awareness would come in locking horns with an institution—the Public Health Service—that was, at least in the public mind, synonymous with scientific objectivity. "The posture of the Public Health Service," Patterson announced, "has been to defend and promote ideas that may be dangerous to the health of all Americans."[60]

As Patterson no doubt understood, the USPHS occupied a special place in the scheme of industrial health. The Lead Industries Association could be expected to defend its members, and even physicians might be granted their self-interest. But the Public Health Service was expected to stand above special interests—to be, in a word, "public." To do so, it had to be independent. In relationship to lead poisoning, this meant standing apart from the firms and industries responsible for lead hazards. And that entailed more than doing research for the lead industries when they wanted it done. It meant defining the problems of lead poisoning before industry defined them; it meant carrying out a program of research that was critical enough to be credible; it meant staffing with personnel who had no previous ties to the lead industry. In all these areas, the Service was deficient, and its product—science—suspect.

Patterson had singled out one agency. But, like others who testified before the subcommittee, he understood that the problem went deeper than any single institution. Quite simply, the lead industries had engineered the development, dissemination, and perception of knowledge concerning the lead hazard. Some of the mechanisms of information engineering, such as the Lead Industries Association, frankly and openly existed to present the lead industry's perspective. Others, including the Kettering Laboratory, translated the industry's needs into the language of science. And still others—semi-public organizations such as the American Public Health Association and the American Medical Association—digested that science and attested to its worthiness.

The result was the suppression of genuine pluralism within the scientific community. Here and there, a dissident voice could be heard. But so complete was industry domination of research into and "knowledge" of the hazards of lead that the central paradigm for understanding lead and its effects remained that pioneered by Kehoe and his associates in the 1920s and 1930s. For the auto industry, the result of this control was almost five decades of unrestricted use of what an Ethyl Gasoline Corporation vice president had in 1925 referred to as the "gift of God"[61] that was tetraethyl lead; for the untold millions who breathed its fumes, it was of course something else again.

NOTES

This essay is adapted from a longer version, originally drafted for The Hastings Center project on occupational health and safety. The project was funded by NSF-EVIST.

1. C. O. Thompson, "Poison in Gasoline," Sept. 8, 1933, statement appearing in the Charleston, SC, *News and Courier,* Record Group 90, "Records of the United States Public Health Service," General Files, 1924–1935; E. Elbridge Morrill, Jr., "Tetraethyl Lead Poisoning Incident with Eight Deaths," *American Industrial Hygiene Association Journal* 21 (December 1960): 515–17.

2. Theodore H. Ingalls, Emil A. Tiboni, and Milton Werrin, "Lead Poisoning in Philadelphia, 1955–1960," *Archives of Environmental Health* 3 (November 1961): 577–79; Eliot Marshall, "EPA May Allow More Lead in Gasoline," *Science* (March 12, 1982):1375–78; *New York Times,* Sept. 7, 1971, p. 17.

3. U.S. Congress, Senate, Committee on Public Works, "Air Pollution—1966," *Hearings Before a Subcommittee on Air and Water Pollution of the Committee on*

Public Works, on S. 3112, A Bill to Amend the Clean Air Act . . . and S. 3400, June 7, 8, 9, 14, and 15, 1966, 89th Cong., 2d sess. (Washington, DC: GPO, 1966), pp. 204–208.

4. *Industrial Medicine* 12 (September 1943): 572; Senate, Committee on Public Works, "Air Pollution—1966," *Hearings,* p. 204; "Symposium on Lead," *Archives of Environmental Health* 8 (February 1964): 13; Joseph C. Robert, *Ethyl: A History of the Corporation and the People Who Made It* (Charlottesville: University Press of Virginia, 1983), p. 119.

5. Willard F. Machle, "Tetra Ethyl Lead Intoxication and Poisoning by Related Compounds of Lead," *Journal of the American Medical Association* 105 (Aug. 24, 1935): 578–85, republished in a collection, University of Cincinnati, College of Medicine, Kettering Laboratory of Applied Physiology, *Experimental Studies on Lead Absorption and Excretion* (Cincinnati, Ohio, 1936), p. 2 of the Kettering collection. This collection is hereafter referred to as Kettering Laboratory, *Experimental Studies.*

6. Zay Jeffries, "Charles Franklin Kettering," National Academy of Sciences, *Biographical Memoirs,* vol. 34 (New York: Columbia University Press, 1960), pp. 107–13; Robert, *Ethyl,* chapter IV.

7. Charles F. Kettering Foundation, *Report for 1961* (Dayton, n.d.), pp. 1 (quotation), 3–5.

8. Robert A. Kehoe, Frederick Thamann, and Jacob Cholak, "On the Normal Absorption and Excretion of Lead," *Journal of Industrial Hygiene* 15 (September 1933): 257, 258, 271 (last quotation), reprinted in Kettering Laboratory, *Experimental Studies.*

9. Robert A. Kehoe, Frederick Thamann, and Jacob Cholak, "On the Normal Absorption and Excretion of Lead, II: Lead Absorption and Lead Excretion in Modern American Life," *Journal of Industrial Hygiene* 15 (September 1933): 282, reprinted in Kettering Laboratory, *Experimental Studies.*

10. Robert A. Kehoe, Frederick Thamann, and Jacob Cholak, "On the Normal Absorption and Excretion of Lead, III: The Sources of Normal Lead Absorption," *Journal of Industrial Hygiene* 15 (September 1933): 296, reprinted in Kettering Laboratory, *Experimental Studies.*

11. Robert A. Kehoe, "Problems in Handling Ethyl Fluid and Ethyl Gasoline," *National Safety News* 29 (January 1934): 19–20, reprinted in Kettering Laboratory, *Experimental Studies.*

12. American Public Health Association, Committee on Lead Poisoning, "Lead Poisoning" (New York: American Public Health Association, 1930), pp. 29, 17, 3; American Public Health Association, Industrial Hygiene Section, "Methods for Determining Lead in Air and in Biological Materials," A Report Prepared by the Subcommittee on Chemical Methods of the Committee on Ventilation and Atmospheric Pollution of the Industrial Hygiene Section of the APHA, 1944 (New York, n.d.); American Standards Association, "American Standard Allowable Concentration of Lead and Certain of Its Organic Compounds," approved Sept. 16, 1943 (New York: American Standards Association, 1943); letter from G. H. Gehrmann, Medical Director, E. I. DuPont, to L. R. Thompson, Asst. Surgeon General, May 29, 1934, in RG 90, USPHS, General Files, 1924–1935, first file in box; and biographical sketch in *Industrial Medicine* 12 (September 1943): 572.

13. Manufacturing Chemists' Association, *Air Pollution Abatement Manual* (Washington, DC: Manufacturing Chemists' Association, Inc., 1952), chapter 1, p. 4 and chapter 4, p. 3.

14. U.S. Department of Health, Education and Welfare, Public Health Service, "Symposium on Environmental Lead Contamination," Dec. 13–15, 1965, sponsored

by the Public Health Service, Public Health Service *Publication* no. 1440, March 1966 (Washington, DC: GPO, n.d.), pp. 155–57.

15. Senate, Committee on Public Works, "Air Pollution—1966," *Hearings,* pp. 205 (quotation), 206, 212, 207, 216 (quotation).

16. Ibid., p. 130; Public Health Service, "Symposium on Environmental Lead Contamination," 1965, p. 147 (quotation).

17. Senate, Committee on Public Works, "Air Pollution—1966," *Hearings,* p. 207.

18. Henry B. Selleck, in collaboration with Alfred H. Whittaker, *Occupational Health in America* (Detroit: Wayne State University Press, 1962), p. 290; U.S., Treasury Department, Public Health Service, "Proceedings of a Conference to Determine Whether or Not There is a Public Health Question in the Manufacture, Distribution, or Use of Tetraethyl Lead Gasoline," Public Health Bulletin no. 158 (Washington: GPO, 1925), pp. v–vii (hereafter referred to as Public Health Service, "Conference on Tetraethyl Lead"); Mellon Institute, "Welcome to Mellon Institute" (Pittsburgh, 1969).

19. William G. Christy, "History of the Air Pollution Control Association," *Journal of the Air Pollution Control Association* 10 (April 1960): 135.

20. Ibid., pp. 131, 135; *Journal of the Air Pollution Control Association* 11 (June 1961): 300.

21. Ludwig Teleky, *History of Factory and Mine Hygiene* (New York: Columbia University Press, 1948), p. 92; "Foreword," *Archives of Environmental Health* 6 (March 1963): 308.

22. *Occupational Medicine* 3 (March 1947): 318–19.

23. Ibid.; Industrial Hygiene Foundation, "History of Industrial Hygiene Foundation" (Pittsburgh: Mellon Institute, 1956), p. 4.

24. John D. Harper, "The Growing Importance of Industrial Hygiene," *Archives of Environmental Health* 6 (March 1963): 315; Industrial Hygiene Foundation, "History of Industrial Hygiene Foundation," p. 19.

25. *Occupational Medicine* 3 (March 1947): 318–19. Others who lent their talents and names to the IHF include Dr. A. J. Lanza, formerly medical director at Metropolitan Life, and C. O. Sappington, a past president of the American Association of Industrial Physicians and Surgeons and the Medical Director of Montgomery Ward & Company in the mid-1920s. See Industrial Hygiene Foundation, "History of Industrial Hygiene Foundation," p. 7, and *Industrial Medicine* 1 (November 1932): 75.

26. *Archives of Environmental Health* 6 (March 1963): 445.

27. Ibid., p. 308.

28. Robert T. P. de Treville, "Natural Occurrence of Lead," in "Symposium on Lead," *Archives of Environmental Health* 8 (February 1964): 212–21; H. H. Schrenk, "Hygienic Lead Standards," *Industrial Medicine and Surgery* 28 (March 1959): 109, 106.

29. "Lead Pipe's Great Record for New Bedford's Water Supply," *Lead* 1 (November 1930): 8; "Use of Tetraethyl Lead Growing Rapidly," ibid.; "Lead Paints Protect Mighty Hudson River Span," *Lead* 1 (January 1931): 4.

30. *Lead* 9 (May 1939): back cover.

31. Joseph C. Aub, Lawrence T. Fairhall, A. S. Minot, and Paul Reznikoff, *Lead Poisoning* (Baltimore: The Williams & Wilkins Company, 1926), p. x; Senate, Committee on Public Works, "Air Pollution—1966," *Hearings,* p. 234.

32. Aub et al., *Lead Poisoning,* pp. 55 (first quotation), 42, 76 (last quotation).

33. Elston L. Belknap, "Differential Diagnosis of Lead Poisoning," *J.A.M.A.* 139 (March 26, 1949): 818–19.

34. Senate, Committee on Public Works, "Air Pollution—1966," *Hearings,* pp. 234–35.

35. Gordon G. Harrold, Stuart F. Meek, "Solubility and Particle Size in Lead Poisoning," in Lead Industries Association, *Proceedings of the Lead Hygiene Conference,* held at Bismarck Hotel, Chicago, IL, Nov. 15–16, 1948 (New York: Lead Industries Association, n.d.), p. 19; F. B. Lanahan, "Preventive Medical Armor for the Lead Industry," ibid., p. 43; Felix E. Wormser, "Facts and Fallacies Concerning Exposure to Lead," in "Conference on Lead Poisoning," papers presented in a Conference on Lead Poisoning at the Seventh Annual Congress on Industrial Health, Boston, Sept. 30, 1946, printed in *Occupational Medicine* 3 (February 1947): 140.

36. *Industrial Medicine and Surgery* 28 (March 1959): 93.

37. Ibid.; Lead Industries Association, *Proceedings of Lead Hygiene Conference,* 1948, pp. 47–48, 53; Philip Drinker, "Public Exposure to Lead," in "Conference on Lead Poisoning," 1946, pp. 145–49, quotation on p. 149.

38. Wormser, "Facts and Fallacies," in "Conference on Lead Poisoning," 1946, p. 135.

39. Senate, Committee on Public Works, "Air Pollution—1966," *Hearings,* p. 233.

40. Ibid., p. 238; Wormser, "Facts and Fallacies," p. 138; Lead Industries Association, *Proceedings of Lead Hygiene Conference,* 1948, p. 8.

41. Lead Industries Association, *Proceedings of Lead Hygiene Conference,* 1948, p. 7.

42. George Rosen, "Patterns of Health Research in the United States, 1900–1960," *Bulletin of the History of Medicine* 39 (May/June 1965): 201–21.

43. Selleck, *Occupational Health,* p. 52.

44. George Martin Kober, "History of Industrial Hygiene and Its Effects on Public Health," in *A Half Century of Public Health,* Jubilee Historical Volume of the American Health Association, ed. Mazyck P. Ravenel (New York: American Public Health Association, 1921), pp. 390, 59 (quotation); New York State, Factory Investigating Commission, *Preliminary Report,* vol. 3 (Albany: The Argus Co., 1912), pp. 1682, 1683, 1685 (testimony of Lawrence Paipken). Reformer John B. Andrews claimed that physicians had falsified death certificates in order to protect a company from liability in two cases of lead poisoning in a smelter. More common, one would imagine, was the simple failure to report lead poisoning to state and municipal health departments, even when required by law. Ibid., vol. 2, pp. 643 (testimony of John B. Andrews, Secretary of the American Association for Labor Legislation) and 465.

45. Selleck, *Occupational Health,* pp. 56ff., 185; *Industrial Medicine* 12 (September 1943): 569.

46. Harry E. Mock, "Industrial Medicine and Surgery—A Resume of Its Development and Scope," *The Journal of Industrial Hygiene* 1 (May 1919): 1–2; C. O. Sappington, "How Far Shall the State Go?" *Industrial Medicine* 1 (November 1932): 75.

47. See the biographical sketch in *Industrial Medicine* 12 (September 1943): 572; Carey P. McCord, "Lead and Lead Poisoning in Early America," Michigan Medical Center, Institute of Industrial Health, Pamphlet No. 6 (Ann Arbor, Michigan, n.d.); "American Industrial Hygiene Association," *Industrial Medicine and Surgery* 28 (September 1959): 407–10.

48. *Industrial Medicine* (Industrial Hygiene Section) 1 (January 1940): 23–24.

49. Selleck, *Occupational Health,* pp. 292–93, 303–304, 316.

50. Ibid., p. 289; American Public Health Association, Committee on Lead Poisoning, "Lead Poisoning" (New York: American Public Health Association, 1930).

51. Ibid., pp. 5, 29 (quotations).

52. American Public Health Association, Industrial Hygiene Section, "Methods for Determining Lead in Air and Biological Materials," A Report Prepared by the Subcommittee on Chemical Methods of the Committee on Ventilation and Atmospheric Pollution of the Industrial Hygiene Section of the APHA, 1944 (New York, n.d.), p. 1.

53. American Medical Association, Council on Industrial Health, *Proceedings of the Seventh Annual Congress on Industrial Health,* Boston, Sept. 30–Oct. 2, 1946 (n.p., n.d.), pp. 7–9.

54. *J.A.M.A.* 104 (Jan. 5, 1935): 200 (Jones), 205 (Kehoe).

55. Ibid., pp. 206, 211; see also Lead Industries Association, *Proceedings of Lead Hygiene Conference,* 1948, p. 8.

56. United States Public Health Service, "The Control of the Lead Hazard in the Storage Battery Industry," from the Division of Industrial Hygiene, National Institute of Health, by Waldemar C. Dreesen, Thomas I. Edwards, Warren H. Reinhart, Richard T. Page, Stewart H. Webster, David W. Armstrong, and R. R. Sayers (Washington, DC: GPO, 1941), p. 1; Hervey B. Elkins, "The Lead Content of Water from Red-Lead Painted Tanks," *Industrial Medicine and Surgery* 28 (March 1959): 112; Lead Industries Association, *Proceedings of Lead Hygiene Conference,* 1948, p. 7. At least one study was done for a trade union.

57. Elkins, "Lead Content," pp. 114–15; Public Health Service, "Symposium on Environmental Lead Contamination," p. 19.

58. Public Health Service, "Control of Lead Hazard," pp. vii, 124 (on storage battery industry); Public Health Service, "Symposium on Environmental Lead Contamination," pp. 30–31; Senate, Committee on Public Works, "Air Pollution—1966," *Hearings,* pp. 130–31.

59. Selleck, *Occupational Health,* the photo section before p. 199; *Archives of Industrial Health* 6 (March 1963): 305–306, 445.

60. Senate, Committee on Public Works, "Air Pollution—1966," *Hearings,* pp. 312–13.

61. Public Health Service, "Conference on Tetraethyl Lead," pp. 115–16; and David Rosner and Gerald Markowitz, "A Gift of God"?: The Public Health Controversy Over Leaded Gasoline during the 1920s," in this volume.

Women, Lead, and Reproductive Hazards: Defining a New Risk

This chapter addresses the health and safety of women in the workplace and demonstrates the significant historical role women have played as both the victims and the major advocates of change. Since few women have been represented in the basic heavy industries (steel, auto, etc.), where the major health and safety efforts have been focused, the jobs that women performed frequently have been ignored. As a consequence of the growth of the service sector and the integration of women into the larger work force, concerns related to the health and safety of working women have begun to receive recognition. The dramatic expansion of the petrochemical industry and the associated production of tens of thousands of new toxic chemicals (about 1000 new chemicals added each year), combined with changing working conditions and relations of production, have made the issue of toxins on the job, at home, and in the neighborhood high priority concerns for workers, consumers, and the general community.

Only when grim and dramatic events occurred resulting from the lack of basic safety precautions in the workplace did women's issues receive widespread attention. When a fire broke out in 1911 in the sweatshop loft of the Triangle Shirt Waist Factory in New York City, one hundred and fifty-four women leaped to their deaths or were burned alive because fire exits were locked. The incident mobilized both public and government officials to institute industrial safety regulations and fire protection programs in New York State—a response after the tragedy, a pattern repeated again and again in the history of occupational health and safety. The incident also spurred on membership in the National Women's Trade Union League, an organization that played a significant role in the fight for better working conditions for women. Similarly, during World War II, when massive numbers of women entered the work force, pointedly in the heavy industries, there was a spate of interest in their particular concerns. When the war was over women were urged to return to their homes, and the millions of women who continued in the work force found their needs ignored once more.

Women scientists have made major contributions to our knowledge of occupational health problems in general, and to the special studies and policy-setting in relation to job-related health issues concerning women in particular. A partial list includes Harriet Hardy, Vilma Hunt, Eula Bingham, and Andrea Hricko. Perhaps the individual who has been most influential in the field has been Alice B. Hamilton. An ardent advocate for protecting workers on the job, her message and loyalties were always clear: "let me beg the industrial physician not to let the atmosphere of the factory befog his view of his special problem. His duty is to the producer, (worker), not to the product."[1] This is a message truly unique for its time, and it remains relevant for contemporary occupational health physicians.

In the early years of the twentieth century, as women's employment in certain industries increased, the need for standards governing working conditions for women began to be articulated. Women leaders, concerned about the plight of women and children as wage earners, called for a government bureau related to the problems of working women. Finally, in 1918, the Women in Industry Service was created; two years later it became the Women's Bureau.

The Bureau recommended a set of standards that sought to regulate all aspects of women's employment. These standards included an eight-hour day, thirty-minute lunch hours, no night work, and the request for "time for recreation, self-development and leisure." The Bureau sought "an adequate wage without discrimination based on sex or race," to "cover the cost of healthful and decent living." In addition to clean, well-lit, and ventilated work sites, it also sought to have all dangerous machinery guarded and all noxious odors and poisons removed. In addition to more mundane recommendations for dressing rooms, rest rooms, and pure water supplies, the Bureau called for "women in supervisory positions . . . where women are employed" and the right of "employees to share in the control of the conditions of employment by means of chosen representatives, not excluding women." The Bureau recommended policies that acknowledged the protection of women and children as a basic societal function, policies that were subsequently implemented in many nations of the world but generally ignored in the United States.[2]

The special protection afforded women workers is rooted in the persisting myth that toxic substances in the workplace influence reproductive capacity only when women are exposed. The disregard for the paternal effect on a broad spectrum of reproductive outcomes including sterility, reduced fertility, miscarriages, low birth weight infants, and the increased mortality of these offspring in the early months of infancy is particularly telling in light of research on the serious consequences of lead exposure in men done in Europe nearly a century ago and similar work done by Alice Hamilton in this country in the early 1900s. In spite of the evidence that lead negatively affects human reproduction in both sexes, the dangers of lead have been viewed largely as a threat to the fetuses of pregnant women. Thus, only women have been singled out for "protection." This has been used as an excuse for excluding women from higher-skilled and higher-paying jobs. The practice of removing "endangered" women workers has created a heated debate over responsibility for risk to the fetus. Underlying the debate is the question of whether the workplace should be made safe or whether "high risk" employees should be excluded from certain jobs.

WHAT ARE THE EXPOSURE/HEALTH EFFECTS?

The knowledge of hazardous materials that Alice Hamilton and other industrial hygienists possessed during the early decades of this century was sufficient to provide methods for prevention of a number of important industrial diseases. Alice Hamilton is best known for her studies of workers exposed to lead, mercury, and common solvents like benzene (benzol).

> We can plead ignorance as to the exact cause of the high sickness rate and death rate of textile workers because we do not know what proportional part is played by the cotton or linen or wool dust, and what part by the fatigue of standing, of noisy, jarring machinery, and what part is played by low wages, by the age or sex or nationality of workers; but when we are asked about the ill health of lead smelters or of workers in benzene coating of artificial leather, we can answer at once that we know the principal cause and that if there are other causes they are only contributory. That is why the study of the poisonous trades, which affect relatively few workers, is highly important because the knowledge is available which is necessary for their control.[3]

Alice Hamilton's most important work was on lead, which she studied so extensively because it was such a common industrial poison, affecting women as well as men. She observed that women in the United States were beginning to enter occupations where exposure to lead was inevitable—the printing trades, glazing of pottery and tiles, manufacturing of porcelain enameled sanitary ware, and the painting trades.

This problem was becoming all the more severe because lead was being introduced in the early 1920s into the larger environment through its use in leaded gasoline. In a journal article in 1925, Hamilton warned that the use of tetraethyl lead would pollute the air and endanger workers who repaired cars.

> I am not one of those who believe that the use of this leaded gasoline can ever be made safe. No lead industry has ever, even under the strictest control, lost all its dangers. Where there is lead, some case of lead poisoning sooner or later develops, even under the strictest supervision, and it is obvious that there cannot be strict supervision of all garages, private and public, all over the country. Thousands of men and boys will be tinkering with motor engines and scattering finely powdered poisonous lead dust about, and it is inconceivable that lead poisoning should not follow. Our best hope is that some non-poisonous substitute for tetra-ethyl lead be found.[4]

In her studies of lead toxicity, she described the deleterious effects of lead on various organs of the human body. But it is her observation and analysis of the effects of lead on offspring that is of particular interest to us here. "Lead is often spoken of as a race poison, in that its effects are not confined to the men and women who are exposed to it in the course of their work, but are passed on to their offspring." Hamilton observed that the effects of lead on male reproduction might be as significant as its effects on females. "The action of maternal plumbism upon the products of conception is clearly demonstrated, but the action of paternal plumbism is less striking. This is of course what one would expect. In the case of the man a poison can act only on the germ cell, but in the case of the woman the toxic action can continue throughout the nine months of pregnancy."[5]

She also pointed out that "there is incontrovertible evidence, both from animal experiments and from observations on human beings, that lead has an injurious action on the *germ cells of both sexes* and that in addition it passes from the maternal to the fetal blood during the months of gestation."[6] (Italics mine.) She specifically studied the effect of lead on women's reproductive functions. "It is a matter of common knowledge among women who work in the lead industries such as pottery and white lead production, that lead has an abortifacient action—this action was noted by medical observers more than a century ago."[7]

Hamilton recognized that lead could affect both sexes, but she was concerned about the repercussions for men, since they were more likely to be exposed to lead than were women. She noted that "the question of whether plumbism in the father may possibly affect the offspring, [is] a question of much greater practical importance to us. . . . there is evidence, based on statistics and confirmed by animal experiments, to show that lead poisoning in the father affects the offspring disastrously."[8] Animal experiments performed at the Universities of Wisconsin and Michigan as early as 1914 had already reported that paternal lead poisoning in guinea pigs resulted in reduction of birth weight, with underweight persisting throughout life.[9] They also showed a very high death rate shortly after birth. This is all quite consistent with what we know of babies who are small or prematurely born and their increased risk of death in early infancy.

Hamilton's early analysis of the reproductive hazard that lead posed to males as well as females is significant because even the impact of maternal factors on reproduction was poorly understood at the time. When maternal German measles infection was first linked with increased complications of pregnancy, birth defects, and low birth weight in the late 1940s, this observation was considered a "new" concept: an infectious agent was transmitted via the mother to the fetus. The potential impact of a chemical on a healthy mother was not clearly understood. It will probably be several decades more before the impact of toxic exposures on the male reproductive system will be thoroughly studied. But for many years the effects of lead on the male system were largely ignored by scientists, health practitioners, and government policy makers—until the episode linking sterility to exposure to the pesticide DBCP in Lathrop, California attracted nationwide attention.

Alice Hamilton recognized that lead was not the only reproductive toxin. She was conscious of the changing nature of industrial production, involving the increasing use of many toxic, synthetic chemicals. In the mid-1920s she expressed concern about the large number of solvents being introduced into production. "We have no figures with regard to the effect of other poisons than lead, but we do know that both carbon monoxide gas and benzol may produce abortion, and that the latter, by causing anemia, renders a healthy pregnancy almost impossible. It is plain to all that if a poison is circulating in the blood of the mother it is practically certain to affect the child she is carrying."[10] She described the serious damage that benzene induces on the blood-forming cells: "This is because chronic benzol poisoning destroys the elements of the blood, causing the victims to suffer not only from a profound anemia but from hemorrhages, for the blood has lost its power to clot . . . If a woman is pregnant, she may suffer a hemorrhage like that of an abortion."[11] In 1933, Hamilton wrote: "I look upon this wide use of benzol as a solvent for the gums and resins as the most serious health hazard that has been developed in industry in recent years."[12]

At that time benzol's ability to drastically reduce the production of red and white blood cells and platelets was well recognized; it was later linked to leukemia. However, it was not until the 1970s that the association was officially acknowledged and the removal of this solvent from industrial settings was encouraged. Consumer products like rubber cement, wood stripping materials, and many others contained benzene. The debate concerning the "safe" level of benzene persists into the 1980s.

SPECIAL STANDARDS FOR WOMEN

In the early years of the twentieth century, the movement among social and labor reformers to provide special protection for women gained strength and legitimacy. A number of states passed protective legislation that limited the number of hours that women could be employed outside the home and controlled certain dangerous aspects of women's work environment. Hamilton was a strong advocate of such measures. But in 1919, she outlined the alternative options available for those seeking to protect women on the job:

> It is very important to look carefully into the question of their employment in such occupations, and to determine whether it will be better to safeguard them by requiring employers to use every known means to reduce or eliminate the hazard of lead poisoning or by prohibiting the employment of women entirely in those occupations in which poisoning constitutes a considerable hazard.[13]

Later, during the New Deal of the 1930s, the issue of special standards for women's employment was evaluated by the Women's Bureau of the U.S. Department of Labor. One of the bases of the Bureau's concern over the effect of the workplace on women was their traditional belief that the special role women played in maintaining the family earned them special consideration. Further, they believed that the earning power of women and the length of their working day were essential determinants of the health and welfare of the society as a whole. But the need for legislation transcended the ideological concern over the maintenance of the existing social order: "Long hours of work and a low wage—the lot of the average woman worker—are great menaces to health. Standards of earnings and working time, such as minimum-wage and maximum-hour legislation, are instrumental in warding off malnutrition and in insuring rest and recreation, thus building up resistance to fatigue."[14]

The horrendous conditions that workers faced during the turbulent years of the Great Depression produced strong support for women's health and safety needs on the job. The Women's Bureau recognized that differential treatment could be used as a wedge separating male and female workers, and therefore specifically sought to defuse this potentially divisive issue:

. . . there is no indication that bad ventilation in a workshop is a more serious menace to women's health than to that of men, nor that it has any distinctive effect on women. Insufficient ventilation will lower the efficiency and the ability to resist disease of both men and women, and it should be recognized as a problem for all employees in all industries under all conditions . . .[15]

They continued by pointing out that their own standards developed to protect women on the job ultimately should be applied to both sexes equally. "In fact, very few if any of these recommended standards can be said to apply only to women, and the Women's Bureau does not advocate that they should be considered as applying to women. The important thing is that they should apply *especially* to women."[16]

The justification for the standards applying especially to women was based not on the concept of the "weaker" sex, but on the full recognition of the double burden that most women bear in the workplace and in the home. "For all conditions in industry bear particularly heavily on women, and therefore good working conditions, hours, and wages have a more important relation to their health. Long hours in the factory are not so serious for the man, who is through work when he leaves his job at night, as they are for the woman, who in many cases has several hours' housework to do after she gets home."[17] The Bureau further pointed out that

the hours of work and the wages paid are factors within the control of management that have a very direct effect on the health of women workers. Most women in industry, unlike most men, have heavy responsibilities outside their working hours. Generally they launder and care for their own clothing and in many cases the clothing of others; large proportions do their own housework and prepare the meals for themselves or a group; where there are children, employed women share in their care or have that entire responsibility. With these facts in mind, the hours spent in work outside the home are seen to form only a part of the working day of women in industry.[18]

In recent years women have found they were being "protected" *out* of their jobs. The issue of protecting women employees on the job and also assuring their right to choose where they work has surfaced as a critical problem. Protection for women often clashes with their right to select desirable jobs. An excellent example of the consequences of female protection are the policies developed by the American Cyanamide Corporation. American Cyanamide developed a "fetus protection policy" whereby women of childbearing age were prevented from holding jobs that the company claimed would expose them to toxic substances at levels that scientists generally believed unsafe for a developing fetus. They maintained that they could not economically (they did not say it was technically impossible) reduce the ambient lead levels sufficiently to eliminate the danger of harm to fetuses carried by women regularly working in leaded atmospheres in the plant.

The company set a policy that barred women between the ages of 16 and 50 from work in high-exposure areas, unless they agreed to undergo voluntary sterilization and the procedure was completed satisfactorily in the opinion of the company's medical advisers. The women who did not agree to the sterilization would have their employment terminated, or would be transferred to other (janitorial) tasks at considerably lower salaries. Five women had themselves sterilized to keep their jobs. The Oil, Chemical, and Atomic Workers International Union (OCAW) fought against the fetus protection policy, claiming that the women had a right to stay on their jobs at their own risk and that of a fetus, should they become pregnant. OCAW argued that the women's right to choose employment was the primary right to be protected.

Under the General Duty clause of the Occupational Safety and Health Act (29 U.S. Code, Section 654),[19] employers must furnish workers with a place of employment free from recognized hazards that cause or are likely to cause death or serious physical harm. Furthermore, the union argued that the fetus protection policy failed to meet this standard, and that workers were entitled to an environment free of serious hazard regardless of the cost to industry. The fetus was to be protected by reducing lead levels, not by excluding fertile women from the workplace.

Given the fact that information documenting the harmful effects of lead exposure on male human reproduction has been available for a long time, the singling out of women for "protection" was particularly inappropriate. Thus, it was only the women who were forced to make a choice between higher-paying jobs and control of their ability to conceive and bear children. Male workers at American Cyanamide remained exposed—their reproductive capacity in jeopardy.

It was not until the decades of the 1960s and the 1970s, a time of increasing activism within the women's and labor movements and heightening concerns regarding health and safety on the job, that the impact of work on childbearing capacity achieved widespread recognition. The research documenting the impact of operating room gases on the pregnancy outcomes of the spouses of male operating room personnel stimulated recognition and research in the realm of the male component of reproduction. However, the work-related sterility problem that dramatized the issue and promoted research and actual policy changes was exposed by the Oil, Chemical, and Atomic Workers Union at the Occidental Chemical plant in Lathrop, California. Its identification of a sterility problem among Agricultural Chemical Division workers influenced the definition of who was at risk and expanded the concept of what was to be considered an occupational disease. Sterility now joined the ranks of recognized occupational health problems along with black lung, silicosis, and certain cancers.

The problem was identified by the workers themselves, who noticed that despite the relative youth of many of the workers in that division and their desire to have children, nobody was having them. When the union brought these issues to management, with specific requests for physical exams that would include sperm testing, the company responded that there was nothing in the plant that could cause those kinds of problems and refused to initiate a testing program. Their attempts through official channels thwarted, a group of workers from the production unit had their sperm tested privately and confirmed scientifically what many of the workers had suspected all along— they were all sterile! The rest of the story unfolded quite dramatically: the subsequent company testing demonstrated that 14 out of the 27 tested had decreased sperm counts, and other cases of sterility were identified throughout the country, all linked to exposure to a chemical called DBCP (dibromochloropropane) which was being formulated in their plant without any warnings or directions for its safe use by employees or precautions concerning known health effects. These circumstances had prevailed until 1977, despite a study published in 1961 by researchers at the Dow Chemical Company and scientists at the University of California School of Medicine, San Francisco. The study described the chemical as "highly toxic on repeated exposure, producing damage at relatively low levels" to vital organs like the kidneys and liver, and to sperm cells and the seminiferous tubules of the testicles.[20] Although the study included information concerning safe handling of this compound, this warning had not been forwarded to producers, formulators, or agricultural and consumer applicators.

The Lathrop plant DBCP exposures were significant for a number of reasons. They illustrated the tremendous toxicity of the newer chemicals generated by the petrochemical industry. They demonstrated once again that often workers are the silent canaries, the human guinea pigs— "identifying" the problems, warning the rest of the society of the problems. Note that it was the workers who identified the problem—not the scientists, the health practitioners, the company, or the governmental regulatory agencies. We are reminded of Alice Hamilton's advice to anticipate the problem, before the problem is demonstrated in the bodies of workers.[21]

The DBCP incident also demonstrated the link between workplace and community exposures. The residues of this chemical were found on many of the crops grown in California. Because of its extensive use in agriculture, it was subsequently identified in wells and ground water serving hundreds of thousands of people. As a consequence of this drinking water contamination, the Environmental Protection Agency (EPA) banned its use in agriculture in 1979. DBCP is still produced in a non-union shop in Los Angeles—then transported across the country in large trucks and shipped out of Gulf Coast ports to Central America. Here it is applied to the crops by unknowing field workers. We then import and consume these fruits and vegetables, thus completing "The Circle of Poison." There are no boundaries or borders for toxic materials.[22]

The Occupational Health and Safety Administration, responding to public outrage and union pressures, acted with rare speed in placing a moratorium on the further production of DBCP. Agribusiness countered with suggestions that perhaps older workers or those interested in sterilization as a method of birth control should volunteer to work with this chemical, as an alternative to banning it. Robert Phillips, the Executive Secretary of the National Peach Council, representing 6300 peach growers in 35 states, chastised OSHA head Eula Bingham and Secretary of Labor Ray Marshall for OSHA's alleged haste in limiting the pesticide's use.

> It appears to us that you (Bingham) and Labor Secretary Marshall may have overreacted, or at least that is your public posture. While involuntary sterility caused by a manufactured chemical may be bad, it is not necessarily so. . . . After all, there are many people now paying to have themselves sterilized to assure they will no longer be able to become parents. . . . If possible sterility is the main problem, couldn't workers who were old enough that they no longer wanted to have children accept such positions voluntarily? Or could workers be advised of the situation, and some might volunteer for such work posts as an alternative to planned surgery for a vasectomy or tubal ligation, or as a means of getting around religious bans on birth control when they want no more children?[23]

OSHA, the Environmental Protection Agency (EPA), and the Food and Drug Administration (FDA) moved to restrict DBCP 16 years after it was linked to sterility in animals. It had been linked to cancer as well.

Tremendous public concern was generated by the revelation that sixteen years earlier a study had been published documenting that DBCP caused testicular changes in test animals, and that workers for years had been continuously exposed to similar levels, resulting in diminished or zero sperm production and the inability to have children. Worker and general public shock and outrage undoubtedly played a role in the initiation of prevention programs. The California State Legislature recognized that the DBCP incident had demonstrated that when vital information about known toxic hazards did not reach the shop floor, serious consequences for public health ensued. Therefore, it created the Hazard Evaluation System and Information Service (HESIS) to provide California workers, employers, and health care providers with the most up-to-date information available on the health effects of toxic substances in the workplace and directions for their safe use.[24] A collect telephone call to a staff of scientists and occupational health practitioners with computer access to appropriate databases could protect future "canaries" from preventable illness and altered functional capacity. The HESIS program was a predecessor of the Worker Right to Know legislation, which was implemented several years later.

Perhaps the most significant outcome of the Lathrop, California Occidental Chemical (a subsidiary of Hooker Chemical) case was the clear identification of sterility as an occupational health problem. Somehow the earlier findings concerning the impact of lead and ionizing radiation had not significantly penetrated the consciousness of either the medical, worker, or general community. After DBCP, hazards to reproduction were finally recognized as a problem for *males* as well as females. In fact, the DBCP incident demonstrated in the workplace what animal tests have established, that rapidly differentiating reproductive tissue (sperm-forming cells) can be particularly sensitive to certain chemicals. This effect is similar to the more generally recognized sensitivity of rapidly differentiating cells during pregnancy. The pronounced effects on sperm formation had occurred at very low levels of exposure. These doses did not produce other clinical signs of toxicity that would be noticed by workers or their physicians, such as damage to the liver, lungs, or kidneys. The warning signs of this occupational "disease" were much subtler than the chronic cough of the miner or the lethal lung cancer of the coke oven worker. "American workers must never be forced to choose between their paycheck and their health."[25]

As a result of the increasing militancy of women and men in the workplace concerning job-related disease and death, and because of the deepening consciousness of reproductive problems, the government finally took action. Dr. Eula Bingham, Assistant Secretary of Labor, issued the first OSHA Standard that intended to set "safe" levels that would protect both male and female workers and their reproductive functions.

In November of 1975, the most comprehensive standard yet issued was unique for several reasons. It was the only standard thus far that took reproductive function and the effects of chronic, low-level exposures into consideration. Most previous standards had been set to prevent the signs and symptoms of acute toxic effects. In the case of lead, the earlier standard was set to prevent the occurrence of convulsions, abdominal colic (symptoms like appendicitis), anemia, and kidney damage; the symptoms of lower-level exposure related to reproductive function, and neurologic effects such as damage to the developing brains of young children, had not been considered. However, perhaps most significant of all from the workers' perspective, the Lead Standard included provisions for on-the-job health screening for lead levels and several specific job protection measures. These measures were to assure that workers could not be "protected" out of their jobs. Workers who were removed from their jobs because continued exposure to lead threatened their health would suffer no loss of earnings, seniority, or other employment rights and benefits during the period of removal. This included the right to be transferred out of the lead area because of concerns related to reproductive function. The standard also required that separate lockers for street clothes and work clothes and showering facilities be provided, to prevent the transfer of lead from the work environment to children in the home.

CONCLUSIONS

What lessons are to be learned from this examination of the history of women and occupational health? Certainly the link between workplace health issues and the health of the community is clearly illustrated by the link between the dual roles of women who work outside the home for wages only to return to the work that must be done at home.

The greater understanding of the nature of women's work in the low paid manufacturing and service sectors, and the increasing involvement of women in the highly toxic "high tech" electronics industry, is helping to focus more attention on the issues of the nature of women's work and associated health problems. Women have played a critical role linking the toxins in the workplace to the increasing burden of toxins in our communities.

The policy debate continues about whether women need special protection. The examination of past and current approaches to protection of reproductive function has illustrated the contradictions inherent in the actual policies. Historically, the need to protect men has been ignored. Those "protections" that were implemented tended to reduce women's job opportunities and excluded them from some of the more skilled and valued male-dominated jobs.

Policies that protect workers should promote job safety and health, not discriminate against a group of workers. Measures that will enhance the nature of work and work performance for women must be available to men as well. At a minimum, these policies should assure that whatever protections are proposed will not deprive women of the same rights and benefits enjoyed by men.

In deciding what jobs women may safely pursue we can pay heed to the wisdom displayed over a half century ago, in a commentary in the *American Journal of Public Health:*

> Women are capable of doing any and all kinds of work. It is not a question of women and industries but of the individual woman and the individual job. . . . It

may come as a blessing to men that permission is given women to work in certain industries, upon the condition that certain health and accident hazards are to be removed. It may humanize industries more. This should not be taken as an argument that men can stand things which women cannot. Men cannot and do not stand them.[26]

The evidence presented on lead and DBCP documents that both men and women are at risk. Until regulatory and employment practices conform to the realities of biology and to the needs of working men and women for a safe and healthful environment for themselves and their offspring, one can anticipate further conflict and tragedy in this area. The options of barring and/or removing women from the workplace, or forcing them to alter their biologic functions, must be replaced with an approach that alters the workplace to conform to human needs and processes. In the short term this means cleaning up the workplace. Hopefully, in the future we will prevent those materials from entering the workplace to begin with. Worker and Community Right to Know laws can help to identify the potential hazards. Appropriate and adequate occupational health and industrial hygiene efforts on the part of unions, employers, and the government must initiate appropriate programs to assure a healthy and safe work place for all workers.

Safe standards of exposure must be set—taking into consideration the impact of chronic, low-level exposures that influence reproduction and the gene pool of our future generations. However, ultimately our society must move toward reducing the production, use, storage, and transport of toxic substances.

The preservation of protective legislation should not be the focus of debate. The issue is *not* whether protective legislation should exist, but how we can effectively extend the protections that women have won to men. The realm where the special needs of women must be addressed relates not solely to paid employment, but to the support services that society must provide to working women and men: adequate, accessible (geographically and cost-wise), quality child care services; paid pregnancy leave to be taken at the mother's and health practitioners' discretion, with job security assured; parental leaves after childbirth; and flexible work schedules with adequate sick leave to meet the needs of care for common (and uncommon) childhood illnesses.

The development and implementation of appropriate protection and prevention programs for women workers cannot be determined by biologic limits, or based on resolving complex scientific issues related to whether male or female germ cells are more vulnerable. Such policies are actually determined by the fundamental social, economic, and political values and related priorities of the society. The basic questions include: Who defines the risks? Who benefits from the activities? Who will pay (literally or symbolically) the costs? Hopefully in the future the burden of working at an unsafe job, or of being selectively protected out of certain types of employment, will not be borne predominantly by women, who still toil long hours outside the home only to return to their second job—caring for children and their homes.

NOTES

1. Alice B. Hamilton, quoted in *Occupational Diseases: A Guide to Their Recognition*, ed. Marcus M. Key (Washington, DC: United States Department of Health, Education and Welfare, 1977).

2. Julia Lathrop, Chair, Committee on Women and Children, Sept. 30, 1919, American Federation of Labor Advisory Conference, NA, RG 86 (Women's Bureau).

3. Alice Hamilton, *Industrial Toxicology* (New York and London: Harper and Brothers Publishers, 1934), p. ix. This volume was part of the Harper's Medical Monograph Series intended for general practitioners.

4. Alice B. Hamilton, "What Price Safety, Tetraethyl Lead Reveals a Flaw in our Defences," *The Survey Mid-Monthly*, 54 (June 15, 1925), 333–34.

5. Alice B. Hamilton, *Industrial Poisons in the United States* (New York: Macmillan, 1925), p. 110.

6. Hamilton, *Industrial Toxicology*, p. 43.

7. Ibid.

8. Ibid. She refers to the work of English researchers Aldridge (1892), Legge (1901), and Oliver (1911), who summarized some of the previous work in the British Medical Journal. These documented the sterility of lead-exposed mothers and the premature deliveries, stillbirths, and mortality rates of babies during the first year of life. See also Hamilton, *Industrial Poisons*, pp. 110–111: "Evidence of the injury to the male germ cell was first adduced by Constantin Paul in 1860, and a number of French authors have since confirmed his statement as to the abnormal prevalence of abortion and stillbirth in the marital history of male lead workers."

9. Hamilton, *Industrial Poisons*, p. 111.

10. Ibid., p. 113.

11. Ibid.

12. Ibid., p. 156.

13. United States Department of Labor, Bureau of Labor Statistics, *Bulletin #253*; Alice B. Hamilton, "Women in the Lead Industries," (Washington, DC: Government Printing Office, 1919), p. 5.

14. United States Department of Labor, Women's Bureau, *Bulletin #136*, Harriet A. Byrne, "The Health and Safety of Women in Industry," (Washington, DC: Government Printing Office, 1935), p. 17.

15. Ibid., p. 4.

16. Ibid.

17. Ibid.

18. Ibid., p. 19; and on p. 4, "The married woman in industry who is forced to work because of economic necessity brought about by her husband's death, incapacity or inability to earn an adequate wage for himself and his family, usually must take whatever job she can get, without too much question of wages or hours. But she is the one worker in all the group who most needs the protection of the law, for the care of her children and household will take many hours and much strength, and her health will suffer if hours of work are not limited."

19. See Occupational Safety and Health Act of 1970 (P.L. 91-596), General Duty Clause, U.S. Code 29, Section 654.

20. T. R. Torkelson, S. E. Sadek, V. K. Rowe, J. K. Kodama, H. H. Anderson, G. S. Loquvam, C. H. Hine, "Toxicologic Investigations of 1.2-dibromochloropropane," *Toxicology and Applied Pharmacology* 3 (1961): 545.

21. Hamilton, "What Price Safety," p. 333: "It makes one hope that the day is not far off when we shall take the next step and investigate a new danger in industry before it is put into use, before any fatal harm has been done to workmen. . . . We hope that some way will be found to institute such inquiries at the beginning, so that after this, animal experiments will precede not follow industrial experiments, and the question will be treated as one belonging to the public health from the very outset, not after its importance has been demonstrated on the bodies of workmen."

22. D. Weir and M. Shapiro, *Circle of Poison: Pesticides and People in a Hungry World* (San Francisco: Institute for Food and Development Policy, 1981).

23. Robert Phillips to Eula Bingham (Director of OSHA), Sept. 12, 1977; cited in *The Guardian* (New York), Oct. 5, 1977.

24. State of California, Department of Health Services, Hazard Evaluation System and Information Service (HESIS), Berkeley, California, 1980.

25. Eula Bingham, "Statement on the Lead Exposure Standard," *Daily Labor Report* (Washington, DC: Bureau of National Affairs, Inc., November 13, 1978).

26. E. R. Hayhurst, "What Industries May Women Engage In?" *American Journal of Public Health* 8 (1918): 861–62.

Part Four

Radium, Asbestos, and Cotton Dust:
Three Paradigmatic Problems of
the Twentieth Century

Throughout American history there have been a number of diseases that were characteristic of their eras. Infectious diseases such as tuberculosis and cholera are closely linked to the growth of the city and urban crowding in the nineteenth century. Today, however, industrial diseases are emblematic of our age. In this section the authors examine three such conditions—radium poisoning, asbestos-related diseases, and brown lung or byssinossis—as a means of unraveling the reasons for growing popular consciousness about these previously undefined illnesses. Angela Nugent looks at the radium dial workers of the 1920s and the conflicting political pressures surrounding scientific discovery of the hazard that affected unskilled, non-unionized working women. David Kotelchuck traces the unfolding of the scandalous story of asbestosis, mesothelioma, and related lung cancers. He analyzes why popular awareness of these conditions only developed in the 1970s although scientists and industry representatives recognized the perils asbestos workers faced. Charles Levenstein, Diane Plantamura, and William Mass look at byssinossis and seek to understand why organized labor was so slow to recognize a disease that affected so many of its members. They point out that in the post-World War II years labor retreated and relegated responsibility for action around health issues to government and professionals. They leave open the questions of what was responsible for the relative inactivity of organized labor: Was it due to regional and ideological factors? Or was it due to the repressive political and social conditions of the postwar years? The three chapters all point to the importance of history in understanding the development of tragedies such as these.

Two of the chapters are rooted in the politically and socially repressive period of the 1920s, when labor was weak and in Calvin Coolidge's phrase the "business of America was business." The story of byssinosis is also rooted in a politically repressive, anti-labor decade, the 1950s. Together the chapters raise an important question regarding the relationship of scandalous health conditions to political and labor repression.

The Power to Define a New Disease: Epidemiological Politics and Radium Poisoning

Among all the occupational hazards created by American industry in the twentieth century, the radiation poisoning of watch dial painters has undoubtedly drawn the most comment and provoked the most study. Contemporaries of the workers poisoned in the 1920s were struck by the dramatic symptoms and fatal course of the disease, and were sympathetic to the class of workers affected—young women whose suffering tapped society's tradition of protective legislation for women and its fascination with the novelty of radioactivity. Historians too have devoted considerable attention to the experience of the radium dial painters. Their story has served variously as a tribute to the medical profession's mastery of a "classic in occupational carcinogenesis," as evidence of the effectiveness of social reform in the 1920s, as an indictment of women's working conditions, and as a somber backdrop to late twentieth-century concern for the hazards of atomic radium.[1]

The episode also suggests a study of the control of information about workers' health, an issue of continuing importance to the movement for occupational safety and health. Legal and scientific conflicts arose over access to data about dial painters' health and working conditions, and over claims of authority to interpret such information. In the course of the controversy, power to define the dial painters' condition shifted from corporate spokesmen to specialists in industrial medicine, hygiene, and radiation. The shift, engineered in great part by the National Consumers' League, confirmed the emergence of a professional community of occupational health and safety experts and society's willingness to cede judgment of workers' health hazards to them. In the case of radium poisoning, scientists' authority to define the disease gave them power to influence workmen's compensation laws, workers' suits for equity, and workplace practices. Their epidemiological research undoubtedly helped to reduce risks for dial painters, but their burgeoning influence muffled public debate and helped to derail a tradition of early twentieth-century lay involvement in decisions about occupational safety and health.

Radium poisoning was a puzzling new disease caused by industrial use of a relatively new chemical. Radioactive elements, still a novelty in the scientific world since their discovery by Marie Curie in 1898, were little-known quantities outside the laboratory. Their luminous characteristics, however, suggested their use in paints, and in 1915 American workers first began using minute amounts of radium and its isotope mesothorium, mixed with other chemicals and adhesives, to illuminate guages and dials. During World War I, workers applied the paint to military instruments, gunsights, airplane dials, and warning signs on military vehicles. After peace was declared, production of luminous consumer timepieces soared to over 2.2 million, with 110 different concerns employing the radioactive paint and two, including the U.S. Radium Corporation, engaging in its manufacture.[2] Far from being perceived as a health hazard, the new technology was employed casually in dial workshops, where managers took no special precautions to protect their workers' health. Indeed, manufacturers promoted the paint as a "new element in the safety movement," arguing that it had the potential to save lives in industry by illuminating power line switches, emergency call bells, and dangerous parts of machines and factory equipment.[3] The medical profession and the patent medicine industry had also advertised the safety of radium. By 1920, radium therapy was well established as a treatment for cancer, and pharmacists sold toothpastes, mouthwashes, ointments, hair tonics, and panaceas that boasted the health benefits of the rare and expensive element.[4]

The existence of a hazard to watch dial painters slowly became evident as an epidemic occurred among the employees of the United States Radium Corporation, a firm which processed radium in West Orange, New Jersey and maintained a staff of 250 women in its watch dial studio. Nine workers died between 1922 and December of 1924 from illnesses marked by severe anemia, lesions on their gums, and necrosis of the jaw. These dramatic symptoms fit no established clinical picture, and the women's death certificates bore a variety of causes of death: ulcerative stomatitis, syphilis, primary anemia, Vicent's angina, phosphorus poisoning, necrosis of the jaw, and "occupational poisoning, character unknown."[5] There had been scattered medical reports of maladies of laboratory personnel who had worked with radioactive elements, but these reports had been overwhelmed by discussion of the therapeutic value of the new substances. Neither the physicians attending the workers who died nor those caring for fifty other employees suffering from similar ailments automatically linked the women's chronic illness to the radiation hazards of their work.

State agencies monitoring workers' health also found that the disease fit no familiar pattern. After being notified of a case of phosphorus poisoning, a compensible disease, in a dial painter, the State Department of Labor investigated the U.S. Radium Corporation, but discovered no evidence of phosphorus and no violations of state law. The department's investigator and consulting chemist voiced strong suspicions of radium, but lacked the mandate to pursue the question of a new occupational disease.[6] The State Board of Health officially viewed the case as probable phosphorus poisoning, an occupational problem conveniently outside its jurisdiction.[7]

While the local community puzzled over the curious cases of sickness among dial painters, specialists in industrial medicine and hygiene learned about the problem in piecemeal fashion. By 1920, workers' health was no longer the province of the loose coalition of trade union members, social justice reformers, efficiency-minded progressives, and public health activists that had sponsored surveys of workers' health before the first World War.[8] New professional associations and training programs for industrial physicians, new hospital clinics for workers with occupational disease, and increased industrial demand for physicians and nurses had created a nationally-recognized cadre of experts in occupational health who were gradually drawn into the controversy over radium.

In the spring of 1924, after one dentist, Theodore Blum, ventured the diagnosis of "radium jaw" in the case of a dial painter with necrosis of the jaw, the U.S. Radium Corporation set out to marshal expert opinion on its own behalf.[9] Although it did not employ an industrial physician, early in 1924 it commissioned the Life Extension Institute, a service which provided medical expertise for industrial firms, to examine volunteer workers. Physicians for the Institute examined six dial painters, and reported that they "showed the ordinary range of human troubles and did not reflect any specific occupational disease," a finding proffered by the company when queried about the health of its work force.[10]

When rumors of industrial hazards persisted, the corporation resolved to hire additional specialists in workers' health to investigate its plant. Arthur Roeder, president of the firm, learned in April 1924 of the field surveys conducted on a contract basis by faculty members of the industrial hygiene program at Harvard, and he requested that Cecil Drinker, a senior member of the staff, study his factory and the health of its dial painters.[11] Drinker assembled a team of physicians who inspected the factory, examined workers, reviewed medical literature on radiation hazards, and collected materials used by the operatives for chemical analysis. From their results, he concluded and reported to Roeder that the dial painters' sickness was occupational in origin and due to radium which operatives inhaled in the form of dust particles and ingested as they pointed the tip of their brushes with their lips—radium which threatened them with the hazards of external gamma radiation. His report argued that the evidence for radium toxicity was so clear "that the burden of proof rests with the person who would maintain that radium has not in some way caused their necrosis of the jaw."[12]

The radium company received Drinker's report, but refused to accept his findings. Arthur Roeder initially answered with a request for additional research, and asked Drinker to carry out experiments which would expose animals to radium and to consider exposure to chemicals other than the mainstay of the firm.[13] Drinker replied with an endorsement of his original report, but obliged Roeder by conducting some of the research he requested. The Harvard physiologist exposed cats to the luminous paint "Undark" in the summer of 1924, and found that radium concentrated in their bones in a manner suggestive of the dial painters' chronic anemias and bone diseases. On receipt of Drinker's new findings, the president of the radium company criticized his work, claiming that it was "preliminary . . . with tentative conclusions," and insisted that the company's view was "that there is nothing injurious anywhere in the works."[14]

Relations between the company and its scientific consultants remained at an impasse until February 1925. When other investigators prepared to report on the dial painters, Cecil Drinker, eager to win recognition for his research, requested permission to publish the results of his team's investigation in a scholarly journal. The company initially ignored the request, and later refused to release the report.[15] Since he had been hired as a private consultant, Drinker considered himself silenced by his contractual obligation to the corporation, but his views changed when he learned that its president, pressed by anxious state officials for information about health conditions at the plant, released a report misrepresenting the findings of the Harvard group. Roeder informed the state Department of Labor that the Drinker's investigation found no health problem among the dial painters and circulated a report which purported to represent the scientists' views.[16] Cecil Drinker, angered by the company's willful attempt to use his reputation to mislead the public, submitted the results of his investigation for publication to the *Journal of Industrial Hygiene* over the protests of the management and lawyers of the firm.[17]

Despite the U.S. Radium Corporation's inability to muffle and control the conclusions of the Harvard specialists in industrial hygiene, the company persisted in its search for experts to establish the safety of its luminous paint. In June 1925, the corporation began aggressive sponsorship of the work of the physiologist Frederick Flinn, a member of the faculty of Columbia University's program in industrial hygiene. The company supplied him with an electrometer, a device to measure radiation contamination, and referred former employees to him for advice on radium's effects. Two dial painters who later died of radium poisoning requested his counsel, and in interviews with each Flinn dismissed fears about the health hazards of radioactive paint.[18] In 1926, after comparing the health of workers at the U.S. Radium Corporation with that of employees at another dial painting firm, the Waterbury Clock Company of New Haven, Connecticut, Frederick Flinn attributed the epidemic in New Jersey to an assortment of communicable diseases. In an article published in the *Journal of the American Medical Association* in December, he absolved the radium painting process and the U.S. Radium Corporation of responsibility for the dial painters' epidemic.[19] Although in his later professional correspondence and publications the Columbia scientist admitted the possibility of radium poisoning, he continued to act as the scientific representative of the corporation, which consistently denied responsibility for their employees' sickness and death. In doing so, Flinn confused public perceptions of radium's dangers and lent credence to the U.S. Radium Corporation's views. He also raised questions in the minds of medical scientists about his ethical standards and susceptibility to corporate influence. Frederick Flinn protested criticisms of his professional behavior, but never explained it to his critics' satisfaction.[20]

From the start of the controversy over dial painting, managers of the U.S. Radium Corporation sought to control information on their workers' health and operated on the principle that expertise in industrial medicine and hygiene was a commercial property to be used to the company's advantage. This strategy delayed recognition of industrial hazards, but ultimately made the corporation vulnerable to workers' legal suits and to public and scientific scrutiny. The New Jersey Consumers' League, a branch of the national association dedicated to the health and welfare of working women, focused on the dial painters' problem, and publicized the U.S. Radium Corporation's attempt to control and distort information about radium poisoning. The league's campaign against the company succeeded in shifting debate over the new disease from the private corporate arena to the courts and a public conference of interested scientists.

The New Jersey Consumers' League was an organization with a local reputation for activism. In the spring of 1924 a health officer in Orange, New Jersey, frustrated by the state's failure to investigate the epidemic among dial workers, informed the league's local secretary, Katherine Wiley, of the deaths and sickness among former employees of the dial-painting firm.[21] Wiley visited the homes of workers and the offices of their doctors and dentists, and, recognizing that a common problem existed, aggressively sought to identify the dial painters' complaint. At first, she queried state agencies, which proved unwilling to conduct further investigations of the local firm.[22] She then sought a study by the United States Bureau of Labor Statistics, whose agents had conducted trailblazing surveys of other trades, but met delays and disinterest on their part. Desperate for information on radium dial painting, she wrote to the Department of Labor and Industries of Massachusetts, only to reach another dead end. The Industrial Hygiene Clinic of the Massachusetts General Hospital, the premier workers' clinic in America, responded but informed her that the dial painters' symptoms were "suggestive of phosphorus poisoning," an occupational poisoning already established as impossible for those who worked with the U.S. Radium Corporation's paints. The company maintained, meanwhile, in correspondence with Wiley, that the dial painters' disease was not occupational in origin.[23] In the face of confusion among local medical practitioners, disinterest and complacency on the part of official agencies, and the intransigence of the U.S. Radium Corporation, Katherine Wiley resolved to arrange for independent analyses of the dial workers' epidemic, and began to monitor the company's handling of the controversy.

In order to implement these goals, Wiley first turned to Frederick Hoffman, a local statistician with a national reputation for research on occupational hazards, and requested that he study the health risks of dial painting. After studying the plant and its workers, Hoffman determined that the unusually high mortality and morbidity among dial painters called for medical investigation, and concluded that some occupational hazard was most likely involved. He reported his finding to Katherine Wiley early in 1925, and formally presented a report to the American Medical Association later in the year. His was the first publication to identify the epidemic among dial painters and to suggest that the hazard lay in the hygienic conditions of their work.[24]

Wiley next sought to follow up on Hoffman's research with a medical study, and welcomed the offer of a medical survey tendered by Alice Hamilton, the industrial toxicologist and a longtime friend of the Consumers' League. Hamilton declared herself "very much interested in these cases," and planned to conduct a full clinical and field study for the league, but she discovered in February 1925 that her colleagues at the Industrial Hygiene Program at Harvard had conducted a study for the corporation, and decided that professional courtesy prevented her from conducting a competing investigation.[25] Independent investigations of dial painting did begin, however, in May 1925, when the Essex County Medical Examiner, Harrison Stanford Martland, performed the first autopsy of an employee of the U.S. Radium Corporation.[26] Martland discovered high concentrations of radioactive substances in the chemist's skeleton, and associated this abnormality with a colleague's query about the clinical symptoms of a less socially prominent employee of the radium firm. Martland initiated an epidemiological study of women dial painters that pieced together evidence from clinical and autopsy studies, and in October 1925 published the first in a long series of articles which documented the hazards of radiation.[27]

After Martland became involved, the Consumers' League turned its attention from sponsoring scientific research toward bringing it into the legal and political arenas. In order to assist a lawyer for the dial workers in assembling a legal case against the U.S. Radium Corporation, Katherine Wiley and Alice Hamilton relayed information about the company's management of Cecil Drinker's report and about Frederick Flinn's erratic activities.[28] In 1927, five injured dial workers, barred from receiving awards from the state workmen's compensation system because of the slow onset of their disease, sued their former employer on the grounds that the company suppressed medical information and directly deceived them about the nature of their illness. The U.S. Radium Corporation, in the face of testimony from Katherine Wiley, Cecil Drinker, and Alice Hamilton, as well as correspondence detailing the controversies over medical research, decided to settle the highly publicized suit out of court, and awarded the five plaintiffs a cash settlement of $10,000 each, an annuity of $600 per year, and medical care for life.[29]

The national office of the Consumers' League also drew on Alice Hamilton for assistance in marshaling experts to call for a national conference on the hazards of dial painting. In May 1928, after three deaths from radium poisoning were reported in Connecticut, Florence Kelley called a meeting of the board of her organization to discuss the necessity of a nationwide study to clarify the nature and the extent of the dangers posed by radium paint.[30] At Kelley's request, Alice Hamilton framed an open letter to Hugh S. Cumming, the Surgeon General of the United States Public Health Service, which was signed by specialists in industrial medicine, medical researchers, and representatives of hospitals and public health agencies in New York, New Jersey, and Connecticut.[31] Their petition for a conference to discuss the poison "whose toxicity and mode of action are still not understood" became part of a carefully orchestrated publicity campaign designed by the Consumers' League to build public support for a national investigation of dial painting. Representatives of the Consumers' League arranged for Walter Lippmann's newspaper, the *New York World*, to publish a series of feature articles and editorials about the dial painting hazard, and they released a copy of the scientists' letter at the climax of public interest.[32]

The campaign of the Consumers' League was successful in achieving its objectives. Under mounting pressure from the general public and the scientific community, Surgeon General Cumming agreed to hold a conference on radium dial painting, and the meeting led to an extensive investigation of the industry by the Public Health Service. By arranging for the conference, the Consumers' League opened up public debate on radium paint and gave specialists in industrial hygiene and medicine a key role in shaping the agenda for discussion and research.

Over the course of the controversy over radium poisoning, the Consumers' League acted as the sole organization representing the dial painters' interests. Its involvement in the issue stemmed from a strong tradition of reform based on cooperation between women workers and middle-class women motivated by concerns for social justice.[33] Individual dial painters confided their problems to a representative of the league, and the organization provided them with support and services no other social group offered. Contemporary labor unions assumed no responsibility for the welfare and safety of the non-union dial painters. The extent of the American Federation of Labor's involvement in their predicament was a representative's token appearance at the Public Health Service conference on radium paint.[34] The Workers' Health Bureau, the independent trade union agency established in the 1920s to study and report on occupational hazards, also took no part in the controversy. Even the United States Department of Labor, the federal agency historically most devoted to the broad problem of workers' health, conducted only belated and peripheral investigations of dial painting. The Consumers' League, by default, became the radium dial painters' chief advocate, and its campaign for public and scientific evaluation of the new and complex poisoning shaped the subsequent history of the disease.

Surgeon General Hugh Cumming may have overstated the case when, in responding to the call for a conference on radium painting, he claimed that the scientists petitioning "should carry enough weight in industrial medicine and research to move almost anything," but medical experts did exert extraordinary influence on the resolution of the radium dial problem.[35] Research scientists used the conference as a forum to wrest from manufacturers the power to determine the occupational hazards posed by radium dial painting. They set the agenda for the conference, established the focus of the investigation that followed, and defined the terms for future discussion of radium's risks.

The conference opened with manufacturers' extravagant professions of deference to specialists in workers' health and the new science of radiation, and closed, to the industrialists' consternation, with those experts determining the meeting's outcome. In the light of public scrutiny, manufacturers found it prudent to appear to defer to scientific expertise. Representatives of the U.S. Radium Corporation attended the meeting and made no protest, and representatives of three watch manufacturers employing radium paint declared their willingness to abandon the use of the process if medical evidence proved it dangerous.[36] Not surprisingly, however, conflicting claims of businessmen and scientists provided the major theme of the meeting.

The first major difference concerned the definition of the occupational hazard involved in dial painting. Businessmen, in defense of their industry, claimed unanimously that the dangers of dial painting resulted from workers' habit of pointing their brushes in their mouths in order to draw delicate lines on watch dials and clocks.[37] Specialists in workers' health were less willing to attribute the risks of the dial painting process entirely to workers' ingestion of the paint. Environmental hazards in the form of airborne radioactive dusts, radon gas pervading the workplace, and atomic radiation from the paints were suspect in the eyes of Cecil Drinker, Alice Hamilton, and John Roach of the New Jersey State Department of Health, who urged, and secured over the objection of manufacturers, investigation of all those potential modes of exposure.[38]

Businessmen and scientists also disagreed over the prospect of radium poisoning outside New Jersey and Connecticut, where the only documented cases had occurred. It seemed reasonable for manufacturers to conclude from past experience refining radium, from dosing themselves with radium nostrums, and from observing workers who applied radium paint that radium was safe to use industrially, since they had personally encountered no cases of radium poisoning.[39] Conference participants with expertise in industrial hygiene were quick to point out the flaws of such common-sense analyses. Ethelbert Stewart of the United States Bureau of Labor Statistics argued that the manufacturers' negative evidence was inconclusive. In his experience, occupational diseases were often mistakenly diagnosed unless specifically studied, and he asserted that no cases of radium poisoning had been recognized in the past simply because the disease was unknown.[40]

From the start of the conference, Alice Hamilton underscored the difficulties of identifying "insidious" chronic cases years after workers' exposure.[41] John Roach added that turnover among workers made occupational diseases especially difficult to trace, and he described the logistical problems of charting the medical history of a mobile work force of young women dial painters in communities more populous and less intimate than West Orange, New Jersey.[42]

Manufacturers who used radium in large quantities were not convinced by the specialists' logic, and sought to circumvent a full-scale study of the dial-painting industry. Over the lunchtime recess they prepared a resolution calling for the surgeon general to appoint committees narrowly charged to establish health standards for future dial painters and to codify protective methods.[43] On hearing the businessmen's proposal, specialists in industrial hygiene rallied to insist on the necessity of field studies of hazards and the need for epidemiological studies of presently employed dial painters; their motion passed the conference by a voice vote.[44]

The most fundamental difference among conference participants involved their attitudes toward assessing the hazards of dial painting. For medical scientists, the issue was primarily a technical one involving the relative merits of different techniques for measuring human radioactivity and the standards for establishing the existence of radium poisoning. For other participants, the issue raised social and moral considerations. Manufacturers reported on the importance of radium to their firms and its usefulness to the military, to passengers in dark railway cars, and to patients in dim sickrooms. They argued that potential industrial hazards must be weighed against radium's benefits, and one clock manufacturer claimed, quite emotionally, that "a radium clock has been a God-send to a great many people."[45] The dissenting view was presented by Ethelbert Stewart, whose moral sensitivity remained keen after forty-one years in the federal service. He protested that radium paint was "purely a fad," and questioned whether there was "enough real utility in the manufacture of luminous watch dials and that sort of thing to pay for what is happening."[46] The debate was forceful but remained unresolved at the close of the conference, and it was the technical experts, with their concern for defining, measuring, and devising controls for hazards, who emerged with the power to investigate and report on the dial painting industry.

In 1929 the United States Public Health Service assembled a team to study the hazards of radium dial painting. Government physicians and a sanitary engineer, supervised by a committee of experts in the field of radiation, investigated seven major dial studios and their workers. The investigators designed their study to determine the nature of the radium hazard, and they established that exposure to airborne radium dust accumulated the element in workers' bodies. Clinical examinations, blood tests, sensitive tests of workers' expired air, and X-ray studies showed that workers who never placed a dial-painting brush in their mouths stored radium in their bones.[47] The investigation documented the existence of an industrial radium hazard outside the studio of the U.S. Radium Corporation, but its authors did not evaluate the risks posed. Instead, it proposed methods to contain them. The government report concluded with recommendations for the radical redesign of dial-painting studios, the introduction of ventilating systems to remove dust and radioactive emanations, provision of lead shielding for workers who measured and mixed the paints, and strict procedures for cleaning and inspecting the workplace.[48] The issue of banning radium paint was never broached.

Evidence suggests that the survey conducted by the Public Health Service, supported by independent clinical and pathological research on the radium hazard, brought about a transformation in the dial-painting industry.[49] With radium poisoning established as an occupational disease and the modes of exposure described, manufacturers adapted their technology to conform to the new knowledge of radioactive hazards. In typical fashion, workers received less protection than consumers, whose radium nostrums disappeared from the market in the panic over the toxicity of the chemical, but dial painters were shielded from the major hazards of the paint.[50] By 1937, workmen's compensation systems in five states included radium poisoning as a compensable disease, and factory inspections by state boards of health and departments of labor checked dial-painting studios for dangerous practices.[51]

By World War II, when industrial demand had increased the number of people working with radium fifty-fold, the dial-painting industry had radically changed. Dial-painting firms of the 1920s had modeled their operations on artists' studios and involved fairly casual management of workers, but the model factory in 1943 patterned itself after a tightly supervised laboratory. Dial painters labored at individual booths, each equipped with ventilation systems, and engineers and inspectors regulated their attire, work routines, and hygiene.[52] The U.S. Bureau of Standards and state governments set standards for establishments using radio-luminous paint, and their rules conformed to the recommendation of the earlier Public Health Service report.[53] Medical management of the workers also assumed an important role. Industrial physicians attempted to contain the hazards of occupational diseases by monitoring workers' health on a periodic basis. At hiring, and twice a year thereafter, workers received physical examinations and tests for radioactivity. If company doctors found them likely candidates for radium poisoning, or if tests showed that they retained more than one-tenth of a microgram of radium, it was general policy to dismiss dial painters or to rotate them to less dangerous jobs.[54] Workers, as a result, had little if any influence over the extent of their exposure to radium hazards.

The controversy over dial painting was a benchmark in American thinking about occupational disease, and its significance is set in relief when it is

compared to an earlier debate over workers' health, the campaign to eliminate the manufacture of white phosphorus matches in the Progressive era.
The diseases caused by exposure to radium and white phosphorus bore a
striking, if superficial, likeness to one another, but the industrial hazards
involved were handled in very different ways. Discussion of phosphorus
poisoning was the province of reform organizations, legislatures, and interested laymen, and centered on the key issue of evaluating the social cost
of one technology for matchmaking.[55] Radium poisoning, in contrast, quickly became the province of medical experts, who addressed their efforts
toward controlling the hazards of radium. As a result, "phossy jaw" from
fabrication of matches disappeared, taxed out of existence by the White
Phosphorus Match Act of 1912, which made white phosphorus matches
prohibitively expensive to buy, while radium poisoning persisted, although
reduced to a minimal level by a medical form of scientific management.

The difference in approach reflected the particularly baffling properties
of radium, the increasing sophistication of industrial medicine, and also a
growing reliance on the judgment of supposedly disinterested experts.
These experts perceived radium poisoning as a problem to solve rather than
to eliminate, and their solution justified their continuing involvement in the
management of dial-painting firms. The technical suggestions proposed by
Public Health Service officers as ways to control radium poisoning became
policies accepted by industry and implemented under the supervision of
industrial health professionals. The judgment of technical experts superseded laymen's open debate about the risks of dial painting after the conference on radium in 1928.

The history of radium dial painting can be charted as the migration of
power over occupational hazards. The major dispute was not a class struggle,
since labor took no part, and the sick and frightened women workers
involved devoted their meagre reserves of militance to suits for equity.
Instead, middle-class women reformers fought for public scientific scrutiny
of the risks of dial painting against the opposition of a corporation that sought
to control and distort information for its own ends. The reformist National
Consumers' League achieved its goals of full access to information about the
hazards of dial painting and improved conditions for workers, but at the cost
of conferring power over industrial hazards to technical experts.

Specialists in radiation, industrial hygiene, and medicine were the
groups that profited most from the controversy over radium dial painting.
These experts appear as far from heroic figures; scientists were slow to
recognize the need for independent investigation of radium poisoning and
devoted much of their initial effort to disputes over professional ethics and
battles for precedence. Nevertheless, they won the power to evaluate the
state of dial workers' health and to prescribe programs for the prevention
and monitoring of future outbreaks. The novel and subtle problem of radium
poisoning appeared in a decade when medical experts had developed effec-

tive techniques to analyze it, and professional reputations strong enough to win public recognition for their views. Rather than serving as technicians who presented information for the consideration of interested parties, they developed into decision-makers whose expertise became compelling. In consequence, they determined the course of occupational health for thousands of workers, and further enhanced their own social status.

N O T E S

The Ellen Swallow Richards Fellowship of the American Association of University Women and a grant from the National Institutes of Health (Grant LM 03785-01) funded research for this paper. An earlier version was delivered in December 1982 at a meeting of the American Historical Association, and I would like to thank Daniel Nelson for his remarks at that time, and John Young for his comments on later drafts.

1. For the medical history, see William D. Sharpe, "The New Jersey Radium Dial Painters; A Classic in Occupational Carcinogenesis," *Bulletin of the History of Medicine* 52 (1978):560–70; and Samuel Berg, *Harrison Stanford Martland, M.D.; The Story of a Physician, a Hospital and an Era* (New York: Vantage Press, 1978). For a celebration of the Consumers' League, see Josephine Goldmark, *Impatient Crusader; Florence Kelley's Life Story* (Urbana: University of Illinois Press, 1953). An interpretation of the episode in terms of women's history appears in Vilma Hunt, "A Brief History of Women Workers and Hazards in the Workplace," *Feminist Studies* 5 (1979):274–83. Histories with an eye toward the A-bomb are: Lawrence Badash, *Radioactivity in America; Growth and Decay of a Science*, (Baltimore: Johns Hopkins University, 1979); Daniel Paul Serwer, "The Rise of Radiation Protection; Science, Medicine and Technology in Society, 1896–1935," (Ph.D. dissertation, Princeton University, 1977); Daniel Long, "A Most Valuable Accident," *New Yorker* 35 (1959):49–87; and Ronald L. Kathren and Paul L. Ziemer, "Introduction: The First Fifty Years of Radiation Protection—A Brief Sketch," in *Health Physics; A Backward Glance*, ed. Ronald L. Kathren and Paul L. Ziemer (New York: Pergamon Press, 1980), pp. 1–8.

2. Serwer, p. 129; and "Survey of Industrial Poisoning from Radioactive Substances," *Monthly Labor Review* 28 (1928):1208, 1219.

3. "Radium is becoming of ordinary household use," *Current Opinion* 69 (1920):537–38.

4. Badash, pp. 130, 131, 147.

5. Harrison S. Martland, "Occupational Poisoning in Manufacture of Luminous Watch Dials," *Journal of the American Medical Association* (Hereafter: *JAMA*) 92 (1929):468.

6. "Survey of Industrial Poisoning," p. 1220; and Lillian Erskine to John Roach, Jan. 25, 1923, evidence submitted in trial, Grace Fryer et al. v. U.S. Radium Corporation, transcript, reel 3, Raymond Berry Microfilms, National Consumers' League Collection, Library of Congress.

7. Charles V. Craster to John Roach, Jan. 3, 1923, in transcript of trial, reel 3, Raymond Berry Microfilms.

8. See Angela Nugent, "Fit for Work: The Introduction of Physical Examinations in Industry," *Bulletin of the History of Medicine 57* (1983):578–83; Daniel Nelson, *Managers and Workers: Origins of the New Factory System in the United States, 1880–1920* (Madison: University of Wisconsin Press, 1975), pp. 117–18; Stuart D. Brandes, *American Welfare Capitalism, 1880–1940* (Chicago: University of Chicago Press, 1976), especially pp. 92–102; Don D. Lescohier, "Working Conditions," in *History of Labor in the United States, 1896–1932*, with an introduction by John R. Commons, vol. 3 (New York: The MacMillan Company, 1935), pp. 316–35; Henry B. Selleck and Alfred H. Whittaker, *Occupational Health in America* (Detroit: Wayne State University Press, 1962).

9. Theodore Blum, "Osteomyelitis of the Mandible and Maxilla," *Journal of the American Dental Association 11* (1924):805.

10. "Survey of Industrial Poisoning," p. 1223; LaPorte vs. U.S. Radium Corporation, 13 Federal Supplement 263, p. 268; typescript, "Report of an investigation of the cause of necrosis of the jaw occurring in certain workers employed by or formerly in the employ of the U.S. Radium Corporation," (Report of Cecil Drinker), p. 15, Transcript of trial, Berry Microfilms.

11. Arthur Roeder to Cecil Drinker, March 12, 1924, reel 1, file 2. Raymond Berry Microfilms.

12. Cecil Drinker to Arthur Roeder, June 3, 1924; and "Report of an investigation of the cause of necrosis of the jaw," Berry Microfilms.

13. Arthur Roeder to Cecil Drinker, June 6, 1924, reel 1, file 2, Berry Microfilms.

14. Cecil Drinker to Arthur Roeder, Nov. 14, 1927; and Arthur Roeder to Cecil Drinker, June 18, 1924, Reel 1, File 2, Berry Microfilms.

15. Cecil Drinker to Arthur Roeder, Feb. 17, 1925; and Arthur Roeder to Cecil Drinker, April 9, 1925, Reel 1, File 2, Berry Microfilms.

16. Alice Hamilton to Katherine Drinker, April 4, 1925; Cecil Drinker to John Roach, April 22, 1925; John Roach to C. Drinker, May 1, 1925; Cecil Drinker to Arthur Roach, May 29, 1925; and Cecil Drinker to Arthur Roeder, June 18, 1925, Reel 1, File 2, Berry Microfilms.

17. William B. Castle, Katherine R. Drinker and Cecil K. Drinker, "Necrosis of the Jaw in Workers Employed in Applying a Luminous Paint Containing Radium," *Journal of Industrial Hygiene* (Hereafter: *JIH*) 7 (1925):378.

18. Deposition of Katherine Schaub, Reel 1, File 1; and C. B. Lee to Frederick Hoffman, Nov. 16, 1926, Reel 2, Berry Microfilms.

19. Frederick B. Flinn, "Radioactive Material an Industrial Hazard?" *JAMA 87* (1926):2078–81.

20. For a review of Flinn's activities see Martland, p. 473. For his defense, see the exchange of letters between Alice Hamilton and Frederick Flinn, November and December 1927, Box 2 Folder 23, Alice Hamilton Collection, Schlesinger Library, Radcliffe College; and Frederick B. Flinn to Harrison Stanford Martland, Feb. 11, 1929, PC/1, Radium 3, Harrison Stanford Martland Collection, George C. Smith Library, College of Medicine and Dentistry of New Jersey, Newark, New Jersey.

21. Lenore Young to Katherine Wiley, April 4, 1924, Box 53 File X-11, Radium Poisoning, National Consumers' League Collection.

22. Katherine Schaub, "Radium," *Survey Graphic 68* (1932):138–40; Report by Katherine Wiley, June 6, 1924, Radium Cases File, Wiley Reports 1924–1928, Papers of the Consumers' League of New Jersey, Archibald Stevens Alexander Library, Rutgers University.

23. Katherine Wiley to John B. Andrews, June 19, 1924, American Association

for Labor Legislation Papers, Catherwood Library, Cornell University; and John W. S. Brady to Katherine Wiley, June 25, 1924, Box 53, File X-10, National Consumers' League.

24. Frederick L. Hoffman, "Radium (Mesothorium) Necrosis," *JAMA* 85 (1925):963–65.

25. See Alice Hamilton to Katherine Wiley, Jan. 30, 1925, and Feb. 7, 1925, Box 53, File X-10, National Consumers' League.

26. Harrison Stanford Martland, "Some Unrecognized Dangers in Use and Handling of Radioactive Substances," *New York Pathological Society*, N.S., 25 (1925):87–92.

27. Harrison S. Martland, Philip Conlon, and Joseph P. Knef, "Some Unrecognized Dangers in the Use and Handling of Radioactive Substances," *JAMA* 85 (1925):1769–76. For a bibliography of Martland's work, see Berg, pp. 215–218.

28. See Alice Hamilton to Katherine Wiley, Feb. 7, 1925, Reel 3; Alice Hamilton to Katherine Drinker, April 4, 1925, Reel 1; Alice Hamilton to Raymond Berry, 5, Jan. 1928, Reel 3, Berry Microfilms.

29. Martland, "Occupational Poisoning," pp. 472; and Stipulation at Chancery Court Pleading, reel 1, Berry Microfilms.

30. Florence Kelley to Alice Hamilton, May 26, 1928, Box 53, Radium File, National Consumers' League Collection.

31. Copy of letter to Hugh S. Cumming, June 16, 1928, Box 53, Radium File, National Consumers' League Collection; for summary of negotiations among signers see Florence Kelley to John B. Andrews, June 25, 1928, Box 53, Radium File, National Consumers' League Collection.

32. "Federal Investigation of Radium Poisoning Asked by Civic Groups," *The World*, July 15, 1925, p. 1.

33. Allis Rosenberg Wolfe, "Women, Consumerism and the National Consumers' League in the Progressive Era, 1900–1923," *Labor History 16* (1975):378–92; and Goldmark, *Impatient Crusader*.

34. Remarks of E. J. Tracy, representative of the American Federation of Labor, Transcript, "Conference on Radium in Industry, Dec. 20, 1928," p. 22, NA, RG 90, 1340/216, Public Health Service Papers.

35. Hugh Cumming to Alice Hamilton, July 18, 1928, Box 53, National Consumers' League.

36. "Conference on Radium in Industry," pp. 12, 28, 30.

37. Ibid., pp. 5, 30, 31.

38. Ibid., pp. 35, 38.

39. Ibid., pp. 12, 18, 19, 20.

40. Ibid., pp. 25, 28.

41. Ibid., pp. 2–3.

42. Ibid., pp. 37.

43. Ibid., pp. 32.

44. Ibid., pp. 39; also see *The New York Times*, Dec. 21, 1928, p. 14.

45. "Conference on Radium in Industry," pp. 28–29, 30.

46. Ibid., pp. 26.

47. Fred L. Knowles et al., "Health Aspects of Radium Dial Painting I; Scope and Findings," *JIH 15* (1933):454.

48. J. J. Bloomfield et al., "Health Aspects of Radium Dial Painting II; Occupational Environment," *JIH 15* (1933):366.

49. For contemporary reviews of the literature, see: Harrison Stanford Martland, "The Occurrence of Malignancy in Radio-active Persons," *The American Journal of Cancer 15* (1931):2435–2516; and Robley D. Evans, "Radium Poisoning: A Review of Present Knowledge," *American Journal of Public Health 23* (1933):1017–23.

50. Elmer H. Eisenhower, "Standardization: Where We Have Been and Where We Are Going," in Kathren and Ziemer, p. 101.

51. "Survey of Industrial Poisoning," p. 1241; *Acts of the 150th Legislature of the State of New Jersey, 1926,* p. 62; New York State Department of Labor, *Annual Report of the Industrial Commissioner,* 1929, p. 1; "Occupational Disease Legislation in the United States," *Bulletin of the U.S. Bureau of Labor Statistics* #625, (Washington, DC: GPO, 1937), pp. 41, 47, 50.

52. Robley D. Evans, "Protection of Radium Dial Workers and Radiologists from Injury by Radium," *JIH* 25 (1943):256–66.

53. Ibid., p. 267; National Bureau of Standards, "Safe Handling of Radio-Active Luminous Compound." *National Bureau of Standards Handbook* H 27, (Washington, DC: GPO, 1941), pp. 1–14.

54. "Safe Handling," pp. 6–7.

55. See Alton R. Lee, "Phossy Jaw: Federal Police Power," *The Historian* 29 (1966):1–21.

Asbestos: "The Funeral Dress of Kings"—and Others

More than six decades have passed since the first published report in 1924 directly linking asbestos exposure to disease.[1] Since that time scientists have periodically reviewed the growing body of literature on the health hazards of asbestos, assessing the scientific evidence of a causal link between asbestos exposure and specific diseases.[2]

During the last decade, paralleling the sharp increase in lawsuits against asbestos companies, scientists and others have begun writing about the state of knowledge of asbestos-related diseases at different times in the past.[3] Implicitly or explicitly they have examined the Watergate question, central to so many of the lawsuits: *What did asbestos companies know about these diseases and when did they know it?* Because those outside the corporations cannot know what was known or said within the inner circles of management, these reviews necessarily had to assess the state of knowledge within the scientific community at various times, and by inference what management knew or should have known about asbestos diseases.

Now, however, lawsuits going on in courtrooms across the United States have unearthed, through the legal discovery process, internal corporate documents that make clear in managers' own words what they knew about asbestos diseases and how they proposed to deal with the related corporate, legal, and human issues. This tale, revealed in bits and pieces through hundreds and thousands of documents, is a tragic one.[4] In effect, the documents reveal the smoking gun of corporate irresponsibility and cover-up by top U.S. corporations in the asbestos industry.

These documents also allow us a much clearer picture—unique in the annals of industrial health in the U.S.—of how one industry established a corporate policy of covering up its products' hazards and how this decision in turn shaped and distorted scientific research and public policy on the issue. The result has been needless loss of tens of thousands of lives over several decades, with many thousands more expected, as well as profound human suffering by the victims of asbestos diseases and their families.[5]

The following review summarizes salient features of the interaction of scientific knowledge, corporate practices, and public policy over the last six decades. Two of its major conclusions are that we urgently need to establish and enforce government standards for limiting workers' exposure to toxic substances on the job, and that we should not entrust this responsibility primarily to the voluntary actions of corporate managers.

ASBESTOS DISEASE AND TECHNOLOGICAL CHANGE IN THE ASBESTOS INDUSTRY

Asbestos—the "magic mineral," the only mineral which can be woven and become a thermal insulator impervious to chemical attack—has been known since antiquity. Pliny described it as "the funeral dress of kings," as well as a potent killer of the slaves who mined and wove it.[6]

Asbestos has been used for centuries to make gloves and other protective clothing for foundry and other furnace workers, as well as resting pads for clothes irons, pots, and other hot objects. In the nineteenth century, after several disastrous fires, it began being used in theatre curtains. Nevertheless it was very costly and its uses limited. In great part the high cost was due to the fact that asbestos was mined by breaking the surrounding rock by metal pick or dynamite and then *handpicking* the individual fibers out of the rock. This laborious method only yielded "long fibers," those longer than about three-quarters of an inch. Thus about 98 percent of the asbestos fiber in the rock was thrown away. Worldwide consumption of asbestos was small, about 500 tons annually in 1880.[7]

Then in the 1890s a major technological breakthrough took place which was to fuel the dramatic growth of the modern asbestos-manufacturing industry over the ensuing decades. Pipe and furnace insulators in the construction industry, looking for materials to give needed structural strength to the well-known insulator magnesia (magnesium silicate), found that addition of 15 percent by weight of short, previously useless asbestos fibers to the magnesia gave a superior insulating product.[8] Patents for this so-called "85 percent magnesia" and similar variants were granted to Keasbey and Mattison in 1886 and H. W. Johns in 1890.[9]

Thus much of the 98 percent of asbestos fiber that was previously discarded now had a major use. For asbestos mine owners, it was as if the yields of their mines had been increased ten-fold or more. Supplies of asbestos, already plentiful, now appeared limitless.

For entrepreneurs like Thomas F. Manville of the newly merged Johns-Manville Company (1901), the task and the opportunity was to buy more asbestos mines and discover ever more uses for the short-fiber asbestos. Soon the company developed or bought rights to processes involving combination of asbestos with wood pulp, to make insulation paper for homes, and with cement, to make asbestos shingles and later transite water pipe. Also during the 1920s, asbestos brake lining came into use for the new high-speed, lead gasoline-fueled cars.

By the mid-1930s, Johns-Manville (J-M) sold a line of 1400 separate products, two thirds of which contained asbestos. In the mines, mechanical rock crushers had moved in to take the place of miners who had picked fibers by hand.[10]

The result of this proliferation of asbestos uses was a dramatic worldwide increase in asbestos production.[11] By 1930, one third of a million metric tons were being produced annually. And this was just the beginning of an upsurge in production that was to reach 4.75 million metric tons in 1974, eighteen times the average annual rate of production during the 1920s (see Table 1).

U.S. domestic consumption of asbestos (most of it mined in Canada and owned by U.S. companies) levelled off during the 1950s and remained fairly constant until 1980 (see Table 2). Since that year, there has been a dramatic drop in consumption, due to the combined effects of the recession in the homebuilding industry and widespread publicity about asbestos hazards.

Throughout these decades medical evidence about the hazards of asbestos exposure was accumulating. The first medical diagnosis of asbestosis, a lung disease caused by breathing asbestos dust, was made by British physician H. Montague Murray in 1900 and reported briefly in his hospital's newsletter.[12] This case was reported to the British parliamentary committee on worker's compensation in 1906, and similar findings appeared in French and Italian publications during the next two years.[13] But these, as well as occasional mentions in the regular annual reports of the British Chief Inspector of Factories, were isolated reports, not widely noted in the U.S. and British medical communities.[14] They would hardly be noted, one assumes, by company president and salesman *par excellence* Thomas F. Manville, who in the words of a 1934 Fortune magazine article "ran the company as [a]

Table 12-1: Worldwide Asbestos Production (Selected Years)

Year	Annual Worldwide Asbestos Production (Metric Tons)	Ratio:	Annual World Production (Col. 2) / Average Annual World Production (1920–1929)*
1924	180,000		0.7
1930	330,000		1.3
1935	420,000		1.6
1942	620,000		2.4
1945	590,000		2.3
1950	1,210,000		4.7
1955	1,910,000		7.3
1960	1,970,000		7.6
1965	2,770,000		10.7
1971	3,770,000		14.5
1976	4,760,000		18.3
1982	3,990,000		15.3

*Average Annual World Production (1920–1929) = 260,000 metric tons
Source: Adapted from *Amer. J. Ind. Med.* 3 (1982): 259–311.

Table 12-2: U.S. Asbestos Consumption (Selected Years)

Year	Annual U.S. Asbestos Consumption (Metric Tons)	Ratio:	Annual U.S. Consumption (Col. 2)
			Average Annual U.S. Consumption (1920–1929)*
1924	170,000		1.1
1930	190,000		1.2
1935	160,000		1.0
1942	390,000		2.4
1945	370,000		2.3
1950	660,000		4.1
1955	710,000		4.4
1960	650,000		4.1
1965	720,000		4.5
1971	690,000		4.3
1976	660,000		4.1
1982	240,000		1.5

*Average Annual U.S. Consumption (1920–1929) = 160,000 metric tons
Source: Adapted from *Amer. J. Ind. Med.* 3 (1982): 259–311.

one-man show. He took no advice, he borrowed no money, he dickered with no competitor."[15]

But if the medical community had paid scant attention to the reports of asbestos-related illness, others were more alert. In a 1918 monograph for the U.S. Bureau of Labor Statistics, Frederick Hoffman, chief statistician for the Prudential Insurance Company, observed that life insurance companies in the U.S. and Canada no longer were selling policies to asbestos workers because of "the assumed health-injurious conditions of the industry."[16] Workers and local management were presumably well aware of this denial of insurance.

1924—ASBESTOSIS
(WORLDWIDE ASBESTOS PRODUCTION—
180,000 METRIC TONS)

In 1924, as noted previously, W. E. Cooke published a case report in the prestigious *British Medical Journal*, clearly linking for the first time asbestos dust exposure and the lung disease asbestosis.[17] During the next few years a number of other case studies were reported, most of them in the British Medical Journal, until by 1930 nineteen fatalities from the disease were reported, including the first reported U.S. case.[18] In most of these cases, the victim had tuberculosis as well, raising the question of whether asbestos dust caused the asbestosis or whether asbestosis was a complication of tuberculosis. This issue was settled in 1928 when a British physician reported a case of asbestosis, confirmed at autopsy, with no signs of tuberculosis—showing that asbestos dust alone caused the disease asbestosis.[19] By 1935, a total of 28 asbestosis fatalities had been reported in Great Britain and the U.S.[20]

In 1924, asbestos-related disease became a major subject of inquiry in the medical community. It was also the last full year of Thomas F. Manville's management of the Johns-Manville Company. During his twenty-four years as company president, Manville oversaw the firm's growth from a small family operation to a large corporation with $40 million in sales annually. There is no indication, to this author's knowledge, of any action taken by Thomas Manville about possible asbestos health hazards.

Manville died in 1925 and left his stock in the company to his brother, Hiram E. Manville; his son, Thomas F. (Tommy) Manville, Jr.; and J-M workers. As a result, J-M employees owned about one third of the company's stock. During the next two years Hiram gained control of the company by buying back large amounts of stock from the workers and from his nephew Tommy. Then in 1927 he in turn sold his controlling share of stock to the financial giant J.P. Morgan and Co.[21]

Morgan immediately brought in Theodore F. Merseles, formerly head of Morgan's Montgomery Ward chain, to become President of Johns-Manville. With regard to the growing number of reports of asbestos health hazards, the company made no public statements, but did begin in 1928 funding animal studies involving asbestos exposure at the Saranac Laboratory of the Trudeau Foundation, a leading U.S. pulmonary disease center in upstate New York.[22]

Merseles, however, died suddenly of a heart attack in March 1929—after only two years as president—and was replaced by his relatively in-experienced personal assistant, Lewis Herold Brown. Brown, then 35 years old, was elected President and Chief Executive Officer of Johns-Manville, posts he held until 1946. At that time he gave up his administrative responsibilities and was elected Chairman of the J-M Board, a post he held until his death in 1951.

During his two-decade tenure, Brown left a strong personal imprint on the company and helped mold it into the corporate giant of today. He protected the health and life of the company during the Great Depression and helped it achieve rapid growth during the 1930s and 1940s.

He was a public relations expert, and he became a national business spokesman.[23] Brown, as we shall see, also instituted the corporate cover-up of asbestos hazards, directed by his brother Vandiver Brown, counsel for Johns-Manville for many years.

1930–35: A CORPORATE COVERUP
BEGINS TO TAKE SHAPE
(WORLDWIDE ASBESTOS PRODUCTION IN 1930—
330,000 METRIC TONS)

After years of case study reports about workers' deaths from asbestosis, Merewether and Price of the British Factory Department inspectorate published an influential government report in 1930 citing asbestos dust exposure as a serious occupational hazard.[24] Soon after, the British Parliament passed new Asbestos Industry Regulations (Statutory Rules and Orders, Nr. 1140, 1931) and made asbestosis a compensible occupational disease.

The Merewether and Price report received much attention in U.S. medical journals. Concerned over the possible impact of these revelations, the Market Analysis Section of the J-M Sales Promotion Department published a survey in 1930 of recent medical reports on asbestos worker deaths for internal J-M use.[25] Thus, at least as early as 1930, J-M officials were well aware of the reported asbestos hazards. In June 1933 the Board of Directors of Johns-Manville approved payment of $35,000 to settle eleven asbestosis suits against them in New Jersey.[26]

On the research front, representatives from Johns-Manville and Raybestos-Manhattan, a major asbestos brake lining firm, asked Dr. Anthony J. Lanza, Assistant Medical Director for the Metropolitan Life Insurance Company, in 1929 to conduct a health hazard evaluation of the asbestos industry. The medical exams and dust counts were completed by January 1931.[27] However, the results were not published until 1935, when they were released as a U.S. Public Health Service Report, with the imprimatur of the federal government.[28]

The results were of course a matter of concern to Johns-Manville and other asbestos companies—however, J-M attorney Vandiver Brown not only reviewed the report before publication, but also actively intervened in the formulation of the article's text and conclusions. This was revealed in the so-called "Sumner Simpson" papers, a series of letters between Brown, Lanza, and Sumner Simpson, then President of Raybestos-Manhattan. The letters were uncovered during the discovery process in a suit against Raybestos-Manhattan, and have since been introduced as evidence in a number of suits against Johns-Manville.

In a 1934 letter J-M lawyer Brown asked Lanza to retain a sentence in the paper's first conclusion that "clinically, from this study it (asbestosis) appeared to be of a type milder than silicosis." He also asked Lanza to re-insert a sentence in the text contrasting the relatively few diagnoses of tuberculosis among asbestos workers to those among workers subject to silicosis. At the same time Brown protested to Lanza, "I am sure that you understand fully that no one in our organization is suggesting for a moment that you alter by one jot or little any scientific facts or inevitable conclusions revealed or justified by your preliminary survey. All we ask is that all of the favorable aspects of the survey be included and that none of the unfavorable be unintentionally pictured in darker tones than the circumstances justify. I feel confident we can depend upon you and Dr. McConnell to give us this 'break.' "[29] Both of the changes recommended above were made in the Public Health Service article, precisely as urged by Johns-Manville.

The results of the study as published showed an epidemic of asbestos disease. The 126 workers examined were selected at random from among active workers, most of them at Johns-Manville plants and mines in the U.S. and Canada. Of these 126 workers, 67 were classified as positive cases of asbestosis based on X-ray examinations, 39 as doubtful, and only 20 as completely free of any X-ray sign of asbestosis.[30] Calculated in percentage terms, these results correspond to a majority of those examined (53 percent) having asbestosis, 84 percent having the disease or some signs on their X-rays (the positive plus doubtful classifications), and only 16 percent with no X-ray signs of the disease at all. These percentages, clearly pointing to the epidemic nature of the disease, were not reported or evaluated in the paper—only the numbers of workers on each classification were used.

Similarly, in worker reports of lung disease symptoms, the authors present numbers of workers reporting symptoms, but do not list the associated percentages—namely that 96 of the 121 queried, 79 *percent*, reported shortness of breath and/or coughing, typical early symptoms of asbestosis. These worker reports, it should be added, were made in an era of job insecurity during the Depression, and before widespread industrial unionization, at a time when ill health often resulted in job loss—see for example the Dreesen study below. Nevertheless, Lanza and his associates summarily dismissed such widespread reports with the comment "Too much emphasis should not be placed on subjective reports of symptoms."[31]

The authors of the study also, at the direct request of the participating asbestos companies, minimized the severity of the disease. As noted previously, they reported no unusual levels of tuberculosis in communities where asbestos was handled, in contrast to the high levels in communities with silicosis disease. Then in their first conclusion, they reported that asbestos dust exposure causes "a pulmonary fibrosis of a type different from silicosis and demonstrable on X-ray films," and then qualified this statement as requested by V. Brown that "clinically, from this study, it appears to be of a type milder than silicosis."[32] These statements were correct reports of the authors' observations in these instances. But instead of emphasizing the widespread prevalence of the disease they uncovered, which they played down in their handling of the percentages of prevalence, the authors emphasized in this first large-scale cross-sectional study of the North American asbestos industry the lesser severity of asbestosis compared to silicosis. A statement of this sort has little content effectively—and is of cold comfort to the victims and their families—if *both* diseases, asbestosis and silicosis, are serious either in terms of severity of illness or mortality.

The authors then went on, inappropriately and improperly, to their fourth conclusion that "asbestosis as observed in this series of cases had not resulted in marked disability in any case."[33] To be sure, the authors did not observe any "marked" disability in this observed series of cases, having dismissed as subjective the symptoms of disability the workers had reported. But their study design, in which the doctors chose to examine only active workers, obviated their seeing cases of serious disability. At the point where workers became seriously disabled, their shortness of breath and general state of weakness presumably prevented them from normal work in an industrial job, and they would have then left the active work force and not been included in the study. Thus, although it was not inappropriate in a first study of the industry to examine the active work force, it was then methodo-logically improper to *conclude*, as the authors did, that the disease did not cause serious disability.

The only part of the study which might have reflected on serious disabil-ity and death among the asbestos workers was an examination of worker insurance claims for death and disability in asbestos companies with group insurance. However, the authors wrote, the numbers of claims was so small that "reliable conclusions cannot be reached."[34] No comments were made by the authors about the social circumstances which might result in fewer or greater numbers of claims, nor was any cohort study undertaken which followed a group of workers forward from some date and continued following them if they left the employ of the company either because of job transfer, disability or death.

This 1935 study commissioned by companies in the U.S. asbestos in-dustry was the first in a long series of industry-sponsored studies which have continued to the present day. Comparison of these studies of various asbes-tos-related diseases with those not funded by the asbestos industry have shown a consistent pattern over the decades of denying that asbestos was the cause of a particular disease, minimization of the severity of an asbestos-caused disease (as in this study of asbestosis), or the shifting of blame for the disease from asbestos exposure to some other cause.[35] Meanwhile, scientific studies not sponsored by the asbestos industry—studies by various govern-ment agencies and by academic-based and personal physicians—just as consistently found asbestos exposure to be harmful to workers' health.

The emerging pattern of active asbestos industry involvement in asbes-tos-disease research and the consistent differences between the results of this and other research becomes clearer with the emergence of another asbestos-related disease—lung cancer.

1935—ASBESTOS EXPOSURE AND LUNG CANCER
(WORLDWIDE ASBESTOS PRODUCTION—
420,000 METRIC TONS)

In 1935, Kenneth Lynch and W. A. Smith, professors at the Medical College of South Carolina, published an autopsy report on an asbestos worker who had both asbestosis and lung cancer.[36] This was notable because at that time widespread cigarette smoking among the U.S. population was in its infancy and cases of lung cancer were rare, especially in combination with asbestosis, another disease which the authors considered unusual.

Other cases soon followed. By 1942, William Hueper stated that the evidence linking asbestosis and lung cancer was "suggestive" that asbestos dust exposure could cause lung cancer.[37] In addition to the active medical research in the U.S. and Great Britain, many German scientists studied and reported a link between asbestos dust exposure and lung cancer.[38]

U.S. asbestos companies, already concerned about reports linking their industry to asbestosis, did not publicly ignore the cancer reports for years, as they had the asbestosis reports. (There was a seven-year time period between the first medical publication in English on asbestosis (Cooke, 1924) and the 1931 publication of the first industry sponsored study, an animal study, by Gardner and Cummings.[39] The time period is at least twice as long if one measures the duration from the time when asbestos was popularly understood to cause lung problems—see e.g., Ref. 24.) An industry-sponsored paper published by Vorwald and Karr of the Saranac Laboratory in 1938 noted the absence of lung cancer among the large group of asbestosis patients treated at that facility and dismissed the various case study reports of lung cancer among asbestosis victims.[40]

Vorwald and Karr argued in their paper that because asbestosis victims were not typical of all asbestos workers—in particular their already diseased lungs might be especially susceptible to lung cancer—a definite asbestos-lung cancer link could not be established based solely on studies of asbestosis victims. What this scientifically correct argument calls, indeed begs, for is an epidemiological study, in particular a long-term cohort study of an initial group of asbestos workers from some particular date of employment over a period of years with all instances and causes of death recorded. But the Catch-22 was that the very asbestos companies which did not want such a link revealed had custody of the personnel records from which such an epidemiological study would be conducted. They did not sponsor this or any other epidemiological study until 1958, twenty years later![41] And the proximal cause of the 1958 industry study was a 1955 epidemiological study by Doll of the prestigious Medical Research Council in England strongly linking lung cancer to dust exposure among asbestos workers.[42] This two-decade period of inaction by asbestos companies, ignoring their own scientific advisors and even excusing their inaction during the war years of 1941–45, shows a callous disregard for the health and lives of their employees.[43] Company records, such as those revealed within the last decade for Johns-Manville, unfortunately confirm that this was the attitude of major asbestos companies.

Under the tenure of Lewis H. Brown, the twenty-year period from 1932 until 1951—the year Brown died—was one of unprecedented growth for the company. During that period U.S. domestic consumption of asbestos rose from a Depression low of 85,000 metric tons in 1932 to 725,000 metric tons in 1951, a spectacular 8½-fold increase in domestic consumption. This record growth accrued overwhelmingly to Johns-Manville, which had an almost monopolistic dominance of the asbestos industry throughout the period.

J-M sales and stock values shot up after 1932. Company sales, which hit a low of $20 million in 1932, had tripled to $60 million by 1937. During this five-year period J-M earnings went from an operating loss of about $1 million in 1932 to a profit of $5.4 million in 1937. J-M stock, down to a low of $10 a share in 1932, was back up to $65 a share in February 1934, already above the value of $50–$55 a share which J. P. Morgan had paid to buy it in the halcyon days of 1927.[44] There was a temporary, one-year fall-off in sales and profits in 1938, the year of so-called Depression II, but sales and profits took off again as World War II approached.

Bringing J-M onto this sales and profits boom was President Lewis H. Brown. The spectacular success of the company he directed, plus his public relations abilities, brought him to national attention as a leader and spokesperson for all U.S. business.

On April 3, 1939, his picture adorned the cover of *Time* magazine, and the cover story was entitled "Public Relations: Corporate Soul."[45] In practice, however, at least in regard to his employees' health and lives, he pursued old-fashioned, single-interest management. He and his brother Vandiver kept a close watch on the asbestos health issue and controlled it as much as they could. As noted above, they had Johns-Manville invest in health research at respected institutions and by capable scientists, but then actively intervened to influence the scientific reports that came out.

Besides intervening in the Lanza paper, Johns-Manville and the other asbestos companies funding animal studies at Saranac Laboratory inserted into their 1936 agreement with Saranac the following provision:

> It is our further understanding that the results obtained will be considered the property of those who are advancing the required funds, who will determine whether, to what extent and in what manner they shall be made public. In the event it is deemed desirable that the results be made public, the manuscript of your study will be submitted to us for approval prior to publication.[46]

Gardner was later criticized by Brown for violating this agreement by referring to the asbestos work in the Lab's 1938 Annual Report and in the Air

Hygiene Foundation Digest. Gardner complained to Dr. Harriet Hardy a few months before his death in 1946 that "Johns-Manville wouldn't allow him to publish his findings."[47]

Asbestos industry companies also acted to keep articles about asbestos health hazards out of their trade magazines. For example, A. S. Rossiter, editor of *Asbestos* magazine, requested in a letter to Sumner Simpson, president of Raybestos-Manhattan, dated September 25, 1935 that he approve publication of a proposed review of asbestosis and control measures. She said in the letter, "Always you have requested that for certain obvious reasons we publish nothing, and naturally your wishes have been" observed. Simpson, in an October 1 cover letter to Vandiver Brown with the Rossiter request, said "I think the less said about the asbestos, the better off we are." In his October 3 letter of reply, Brown said "I quite agree with you that our interests are best served by having asbestosis receive a minimum of publicity."[48] Permission was denied.

This policy of limiting public information about asbestos disease was carried even further with workers who were victims of these diseases. In some cases during this era worker/victims of asbestosis appear to have been summarily fired because of their work-induced disability. For example, when federal industrial hygienists were summoned by North Carolina state officials to conduct health studies on workers in three local asbestos textile plants in 1938, they found that approximately 150 older workers out of a total work force of 540 persons had recently been fired and replaced by younger workers with little or no asbestos experience.[49] Of the 69 fired employees located by the federal scientists, 43 were found to have asbestosis. This was confirmed in several follow-up studies by local physicians.[50]

Interviews with asbestos workers and union officials such as those at the J-M plant in Manville, New Jersey indicate that J-M workers were not told by the company of asbestos hazards or of the results of company medical examinations until the 1970s.[51] This policy is well-documented in many asbestos company records after World War II, but is not so well-documented during the prewar period. However, a 1984 court deposition by Charles H. Roemer, former Chairman of the Paterson, New Jersey Industrial Commission, gives chilling testimony to management attitudes at Johns-Manville during the early World War II period.

Roemer described an invitation sometime in 1942 or 1943 to lunch with J-M president Lewis Brown and his brother Vandiver to discuss how J-M handled its asbestos health problems. Roemer then testified:

> I'll never forget, I turned to Mr. Brown, one of the Browns made this crack (that Unarco managers were a bunch of fools for notifying employees who had asbestosis), and I said, "Mr. Brown, do you mean to tell me you would let them work until they dropped dead?" He said, "Yes. We save a lot of money that way."[52]

During this five-year period there were few major new scientific papers on asbestos-related diseases. However, company doctors in asbestos plants around the country continued to take X-rays and examine asbestos workers, and they had to deal with growing numbers of asbestosis victims employed in the plants.

In 1949 Dr. Kenneth W. Smith, Corporate Medical Director for Johns-Manville, summed up J-M's corporate medical strategy in a now famous memo to his corporate superiors:

> It must be remembered that although these men have the X-ray evidence of asbestosis, they are working today and definitely are not disabled from asbestosis. They have not been told of this diagnosis for it is felt that as long as the man feels well, is happy at home and at work, and his physical condition remains good, nothing should be said. When he becomes disabled and sick, then the diagnosis should be made and the claim submitted by the Company. The fibrosis of this disease is irreversible and permanent so that eventually compensation will be paid to each of these men. But as long as the man is not disabled it is felt that he should not be told of his condition so that he can live and work in peace and the Company can benefit by his many years of experience. Should the man be told of his condition today there is a very definite possibility that he would become mentally and physically ill, simply through the knowledge that he has asbestosis.[53]

How this strategy was manifest is seen in reports of a series of medical conferences about individual workers by J-M medical staff, released and made public during legal suits by asbestos victims.[54] Basically, as the medical conferences indicate, company physicians initially did *not* tell workers that they had asbestos disease. But as the disease progressed through continued exposure and the worker became disabled, the company physician would reveal to the victim—slowly and in guarded terms—his or her true condition. In the following medical case reports, the first date indicating some chest X-ray abnormality in the patient that triggered physician concern is indicated by an "N.D.," a "No Dust" restriction, by means of which these physicians advised the company management to place the worker in a non-dusty environment. The health counseling session ("H.C.") was the time when the physician had a personal meeting with the worker/patient to review his or her medical situation, to present the medical diagnosis and advise on job or personal medical precautions (e.g., give up smoking). Often several years elapsed between this No Dust restriction and the Health Counseling session. Sometimes, as the record below indicates, some mention was made of the person's medical condition short of a full counseling session. (All italicized sentences below are emphases by the author) Consider the following case from the March 5, 1958 medical conference:[55]

Patient: Male, 50 years old.
Dr.: Hired in 1925. Pipe machine operator, 22 years exposure to silica, cement,

and asbestos in transite pipe. Carpenter for 4 years.
Nurse: *N.D. in 1948. Pneumoconiosis mentioned in 1954. H.C. in 1956 of X-ray changes.*
Diagnosis: Early to moderate mixed pneumoconiosis . . .
Nurse: Should these men be advised?
Dr.: *We can put transite pipe out of business from this list alone.*

We see above that six years elapsed between the physicians' No Dust recommendation to the company in 1948 and the "mention" of pneumoconiosis to the victim in 1954. Another two years elapsed before a full health counseling session was given.

As indicated in the last comment, these physicians were well aware of the extent and severity of the dust disease problem, at least in the transite pipe division (where asbestos dust is added to a cement mixture to make a very strong durable water pipe).

The long delay between the medical observation of X-ray changes and notification of the workers was not exceptional; it was common practice. During the four reported medical conferences, a total of 20 workers had the dates both of the No Dust restriction and of their Health Counseling session recorded on their medical report.

For almost two thirds of the workers (13 cases), the health counseling session took place more than one year after the company doctors had recommended that they be placed on a no-dust restriction. For these 13 cases, *the average delay between the job restriction and the worker health counseling was 3.6 years,* with delays ranging from 2 to 10 years![56]

Often when the affected workers were transferred from dusty to less dusty jobs, they were not told that this was done by the company for medical reasons. For example, Joseph Kiewleski, an asbestosis victim, charged that he was transferred from his job as a machinist to a janitor's position soon after a company physical. Years later he found out from his personal physician that the reason for the job change was to remove him from asbestos exposure.[57]

In some cases physicians simply decided not to tell workers of known or suspected medical conditions. For example, in the following case doctors observed possible lung tumors in the X-rays and yet decided explicitly not to tell the affected workers. This medical conference occurred July 10, 1957:

Patient: Male, 50 years old. Hired in 1932. 'Minimal' exposure to silica, asbestos fibre, Portland Cement—12 years.
X-ray: Equivocal area of increased density right apex . . .
Diagnosis: *Might be infection or tumor.* No occupational disease or TB. X-ray changes of unknown origin.
Dr.: 1. No tab [No marker on record—D.K.]
2. No AHS [No Air Hygiene Survey—D.K.]
3. Do not notify plant manager
4. *No H.C.*

In this case, the J-M doctors actively suspected an infection or a tumor, yet they did not tell the patient of their suspicions or take protective actions of any sort. Thus, the worker was not told of a potentially grave condition, so that he might consult other doctors, have other tests performed and possibly have some medical action taken, such as surgery. And since the worker was not informed of his condition, he could not choose to quit, seek to transfer from his dusty job, stop smoking (if he was a smoker), take any other protective measure, or sue for damages.

Another serious case, involving a decision not to transfer a worker with an advanced case of asbestosis (pneumoconiosis), was discussed between Medical Director Smith and two other J-M physicians on March 5, 1958.[58] It was resolved in what can only be called cold-blooded terms:

Patient: Male, 52 years old . . .
Dr. Smith: *Advanced Pneumoconiosis*.
Dr. 1: Should we change him?
Dr. Smith: Won't make any difference.
Dr. 2: If he hits 65 I will be surprised.
Dr. 1: He is to be watched carefully and retire on disability, if necessary.
1. Tab
2. Notify plant manager
3. *Do not transfer*
4. Watch carefully
5. Retire if necessary.

It was not until the early 1960s that Dr. Irving Selikoff and his associates at Mount Sinai Medical Center in New York broke industry's hegemony over medical and personnel information by using the welfare and retirement records of the asbestos insulators' union as the basis for conducting an epidemiological study. For the first time in the United States, scientists not beholden to industry conducted large-scale definitive studies on groups of asbestos workers. Over the course of the next twenty years, a battle would rage over the legal, economic, and ethical consequences of the industry's irresponsible actions. The lesson is especially important in the present era of government deregulation. If we do not learn from these lessons of our recent past, we and our children will be condemned to relive them not as farce, but as continuing tragedy.

N O T E S

1. W. E. Cooke, "Fibrosis of the Lungs Due to the Inhalation of Asbestos Dust," *British Medical Journal* 2(1924):147.

2. E. R. A. Mereweather and C. W. Price, *Report of Effects of Asbestos Dust on*

the Lungs and Dust Suppression in the Asbestos Industry (London: HM Stationery Office, 1930); W. C. Hueper, *Occupational Tumors and Allied Diseases* (Springfield, IL: Chas. C. Thomas Publishers, 1942); NIOSH, *Criteria for a Recommended Standard: Occupational Exposure to Asbestos,* Publication no. HSM-10267 (Washington, DC: GPO, 1972); NIOSH, *Revised Recommended Asbestos Standard,* Publication no. DHEW(NIOSH) 77-169 (Washington, DC: GPO, 1976); I. J. Selikoff and D. H. K. Lee, *Asbestos Disease* (New York: Academic Press, 1978).

3. David Kotelchuck, "Asbestos Research," *Health/PAC Bulletin* 61(Nov./Dec. 1974):1–6, 20–27; Paul Brodeur, *Expendable Americans* (New York: Viking Press, 1974); Daniel Berman, *Death on the Job* (New York: Monthly Review Press, 1974); Samuel Epstein, *The Politics of Cancer* (San Francisco: Sierra Club, 1978), pp. 83–100.

4. Barry T. Castleman, *Asbestos: Medical and Legal Aspects.* (New York: Harcourt Brace Jovanovich, 1984).

5. I. H. Selikoff, *Disability Compensation for Asbestos-Associated Diseases in the United States* (New York: Mt. Sinai School of Medicine, 1982), chapter 8.

6. Pliny, "The Funeral Dress of Kings," quoted in Donald Hunter, *Diseases of the Occupations,* 6th Edition (London: Hedder and Stoughton, 1978).

7. "Management by Morgan," *Fortune* 9, no. 3(April 1934):132.

8. Ibid., pp. 135–36.

9. Stephen L. Berger, "Alternatives to Asbestos Insulation," in Castleman, chapter 6.

10. "Management by Morgan," p. 132.

11. Tables 1 and 2 adapted from W. J. Nicholson, G. Perkel, and I. J. Selikoff, "Occupational Exposure to Asbestos: Population at Risk and Projected Mortality—1980–2030," *Amer. J. Ind. Med.* 3(1982):259–311.

12. H. M. Murray, *Charing Cross Hospital Gazette* (London, 1900).

13. Reviewed in Castleman, pp. 2–6.

14. Selikoff and Lee, pp. 20–25.

15. "Management by Morgan," pp. 83–84.

16. F. L. Hoffman, "Mortality from Respiratory Diseases in the Dusty Trades (Inorganic Dusts)," Bulletin no. 231 (Washington, DC: U.S. Bureau of Labor Statistics, 1918), reported in Castleman, p. 6; and NIOSH, *Criteria for a Recommended Standard,* pp. 3–4.

17. Cooke, p. 147.

18. D. S. Egbert, "Pulmonary Asbestosis," *American Review Tuberculosis* 32(1935):25–34. For a review of these papers, see Kotelchuck, "Asbestos Research."

19. H. E. Seiler, "A Case of Pneumoconiosis," *British Medical Journal* 2(1928):982.

20. Egbert, "Pulmonary Asbestosis."

21. Richard Goodwin (then president of Johns-Manville), in a speech to the Newcomen Society, New York City, Dec. 16, 1971. See also "Management by Morgan."

22. "History of Johns-Manville Health Research" (Johns-Manville Co., mimeographed, 1972), an unpublished manuscript prepared for the 1972 U.S. Labor Department hearings on a permanent asbestos standard and later distributed by the U.S. Asbestos Information Association.

23. "Public Relations: Corporate Soul," (cover story) *Time* (April 3, 1939):52–58.

24. Mereweather and Price, *Report of Effects.*

25. *Pulmonary Asbestosis,* Memo Report no. 22, Market Analysis Section, Sales Promotion Department, Johns-Manville Corp. (March 21, 1930), reported in Castleman, p. 9.

26. Minutes, Board of Directors Meeting, Johns-Manville Corp., June 26, 1933.

27. Castleman, pp. 123–26.

28. A. J. Lanza, V. J. McConnell, and J. W. Fehnel, "Effects of Inhalation of Asbestos Dust on the Lungs of Asbestos Workers," *U.S. Public Health Reports* 50, no. 1(1935):1–12.

29. Letter from Vandiver Brown to A. J. Lanza, Dec. 21, 1934.

30. Lanza et al., "Effects of Inhalation."

31. Ibid., p. 8.

32. Ibid., p. 10.

33. Ibid., p. 11.

34. Ibid., p. 9.

35. Kotelchuck, "Asbestos Research."

36. K. M. Lynch and W. A. Smith, "Pulmonary Asbestosis," *Amer. J. Cancer* 24(1935):56–64.

37. Hueper, pp. 399–405.

38. Reviewed in Castleman, chapter 2.

39. L. U. Gardner and D. E. Cummings, "Studies on Experimental Pneumokoniosis," *J. Ind. Hyg.* 13(1931):97–114.

40. A. J. Vorwald and J. W. Karr, "Pneumoconiosis and Pulmonary Carcinoma," *Amer. J. Pathology* 14(1938):49–57.

41. D. C. Braun and T. D. Truan, "An Epidemiological Study of Lung Cancer in Asbestos Miners," *Arch. Ind. Health* 17(1958):.634–52.

42. "Management by Morgan," p. 82.

43. Raybestos asbestos sales vs. J-M's.

44. "Public Relations: Corporate Soul," pp. 56–57; "Management by Morgan," p. 82.

45. "Public Relations: Corporate Soul," pp. 52–58.

46. Letter from V. Brown to L. V. Gardner, Director of the Saranac Laboratory (Nov. 20, 1936).

47. Letters from V. Brown to S. Simpson (May 3 and 10, 1936) and H. Hardy, quoted in *Washington Post* (Nov. 12, 1978), in Castleman, pp. 48–49.

48. These letters are reprinted in full in Castleman, Appendix 1, pp. 555–58.

49. R. R. Sayers and W. C. Dreesen, "Asbestosis," *Amer. J. Public Health* 29(1939):208.

50. Reported and documented in Castleman, pp. 199–200.

51. D. Kotelchuck and M. Handelman, "Corporate Cancer," *Health/PAC Bulletin* 50(March 1973):6.

52. Deposition by Charles H. Roemer taken April 25, 1984 in *Johns-Manville Corp. et al. v. United States of America*, U.S. Claims Court Civ. no. 465-83C, in Castleman, p. 401.

53. I. J. Selikoff, *Disability Compensation*, p. 30 and Figures 2–5 and 2–6; Kenneth W. Smith, "Industrial Hygiene—Survey of Men in Dusty Areas," J-M internal memo, February 1949.

54. D. Kotelchuck, "Losing Patience: A Look Back at Corporate Medicine in the Asbestos Industry," *Health/PAC Bulletin* 11, no. 5 (May/June 1980):1–2 and 7–14.

55. Ibid., p. 8.

56. Ibid., pp. 9 and 12.

57. Kotelchuck and Handelman, "Corporate Cancer."

58. Kotelchuck, "Losing Patience."

CHAPTER **13**   *Charles Levenstein,*
Dianne Plantamura,
and William Mass

Labor and Byssinosis, 1941–1969

As early as 1941, the cotton mill workers' lung disease—byssinosis or "brown
lung"—was made compensable in Great Britain.[1] Not until 1969, almost
thirty years later, was byssinosis recognized in the United States as an
occupational disease worthy of concern.[2] While the disease may not have
been as prevalent in the United States as in England, ample evidence exists
that American cotton mill workers suffered its effects in substantial
numbers.[3] What accounts for this lag in science and this lack of attention to
workers' health in the United States? Perhaps even more fundamental, why
was brown lung *not* a major issue for the organized labor movement until the
late 1960s? We believe the answers to these questions lie in two related
areas: the weakness of the American trade union movement and the historic
relationship between labor and science.

Section I of this chapter briefly reviews early labor reform concerning
textile workers in order to set the context for Section II's discussion of labor
relations and the economics of the textile industry. Section III shows the
results of our review of official textile labor newspapers and convention
proceedings from 1939 to 1949, including a period of some attention to
cotton dust problems. Section IV describes the decline of textile unionism
since World War II and continues the review of scarce official material on
health and safety until 1968–69. The final section presents conclusions and
speculations about labor's apparent lack of concern with occupational health;
the nature of the health problem vis-à-vis other labor issues; and the politics
of labor and occupational health.

EARLY LABOR REFORM

The problem of cotton dust in the mill is not a new one. A Pittsburgh
physician, testifying before the Pennsylvania Senate in 1837, reported on
conditions in cotton mills:

> The factories are ill ventilated; their atmosphere is constantly impregnated and
> highly surcharged with the most offensive effluvia—arising from the persons of
> the inmates, and the rancid oils applied to the machinery. . . . In the rooms
> where the cotton wool undergoes the first process of carding and breaking, the
> atmosphere is one floating mass of cotton particles, which none but those
> accustomed to it, can breathe, for an hour together, without being nearly
> suffocated.[4]

In the nineteenth century, the efforts of labor organizations and labor reform advocates to improve the situation, however, focused on the reduction of working hours and the removal of people perceived to be particularly vulnerable to occupational hazards—children, but also, in some cases, women. Efforts to control certain safety hazards by requiring machine guarding were also successfully legislated, but by and large management maintained effective control over the technology of production, and their right to do so was seldom challenged. In the early twentieth century, trade unions and public health advocates joined forces to ban the "Kiss of Death" shuttle in Massachusetts, but this was a rare exception to the rule. A more significant legislative thrust, thought to have a preventive effect, was the successful effort to pass workers' compensation legislation. No-fault workers' compensation insurance, which barred worker suits against employers, was the program of major industry groups which won the day.[5] Thus, the prevailing approach was to remove the susceptible and to compensate the irreducible minimum hurt at work, rather than to regulate hazardous technology.

"Ordinary diseases of life," however, were not "occupational" and were not included in state compensation schemes. Textile worker lung disease was common. The usual state of health for textile workers was not very good, and tuberculosis was rampant. Textile workers recognized their dust-related occupational diseases; they named them grinder's asthma, Monday morning feeling, and other similar names for byssinosis, and were forced to accept them as the inevitable facts of life. In any case, the central feature of working-class life was the family, not the individual, and the well-being of the family as a whole had to be maintained. (Exactly who was in the family may have changed over the years, but it was the health of that unit that was of concern.) Maintaining family income was the highest priority in protecting the family's "health." Occupational health problems, as middle-class professionals defined them, were of some importance, but came lower in the hierarchy of concerns of the working poor. One textile worker, speaking decades later, summarized the enduring problem:

> But face it, the money is the most important thing. It's a living. It represents your home and your food and your clothes and everything. And when you're not trained for nothing else . . . even if you were trained the situation's not available. And more or less you just have to go on.[6]

LABOR RELATIONS AND ECONOMICS OF THE TEXTILE INDUSTRY

Textile worker organization goes back as far as 1834, when 2,000 female operatives struck in Lowell over a wage cut, but not until over 100 years later was a successful national union launched with the Congress of In-

Table 13-1   Textile Union Membership, 1897–1975

Year	UTW	TWUA
1887	2,700	
1904	10,500	
1912	10,900	
1920	104,900	
1924	30,000	
1930	30,000	
1935	79,200	
1939	1,300	83,300
1945	43,100	247,600
1955	48,100	197,500
1965	36,000	123,000
1975	36,000	105,000

Source: Appendix 5, "Membership (Selected Unions)," in *Labor Unions,*
ed. Gary M. Fink (Westport, Conn: Greenwood Press, 1977), p. 495.

dustrial Organization's establishment of the Textile Workers Organizing Committee.[7] By the turn of the century, however, craft unions were well established in New England, and especially in Fall River, where in 1899 16 percent of the looms in the country were located. The Fall River unions were the largest and the strongest in the industry; their agreements set the pattern for wage settlements.[8]

The last direct attempt by Fall River manufacturers to break the unions was in 1904–05 during a strike precipitated by severe wage cuts. After six months, the Governor arbitrated a settlement which established low pay rates but reinforced the system of bargaining, and the manufacturers were not able to achieve the even lower Southern wage scales.

Most Fall River manufacturers prospered for a time by turning to specialized orders. Southern mills were offering increased competition on staple print cloth by using Northrop looms, but the large investment required for retooling with Northrops was less attractive in Fall River, where proximity and regular relations with New York cloth markets gave higher prices and an advantage in specialized orders. The labor savings of Northrops were not as great when frequent machine adjustments and greater attention were required because of constant changes in the style and grade of cloth produced. Certain Fall River firms, however, did convert to Northrop looms during the period 1909–1914. The leading textile firm, Fall River Iron Works, was also the first and one of the few local concerns to relocate production to the South. The other manufacturers as a group were hit first and hardest by the post-World War I cotton textile depression.[9]

The textile industry was plagued by a boom-and-bust business cycle

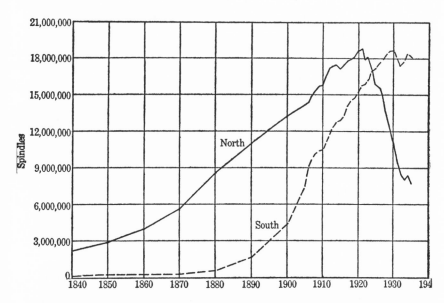

Fig. 13-1 Production Capacity—North and South Active Spindles
Source: Stephen Jay Kennedy, *Profits and Losses in Textiles* (New York: Harper, 1936), p. 199.

which exacerbated trade union organizing difficulties prior to passage of the National Labor Relations Act. The strength that had developed in the Northeast was devastated by the decline in the 1920s that turned into a massive industrial shift to the South. At its height, no more than 10 percent of Southern textile workers were organized.

The causes of the decline in the New England textile industry, the significance of its technological backwardness compared with the rising Southern industry, and the impact of union bargaining on the choice of technology have all been subjects of heated debate.[10] What is incontrovertible is the impact on unionization. Table 13-1 shows the membership figures for the two major textile labor organizations. The Textile Workers Organizing Committee was formed by the CIO in the late 1930s, so the numbers of significance for us are for the United Textile Workers between 1920 and 1930. The decline from 104,900 to 30,000 shows the decline of Northern textiles. Figure 13-1 shows production capacity in textiles, illustrating the dramatic shift of the industry that occurred prior to 1930. The shift of production from Northern mills to the South is evident, as Southern capacity continued to grow until 1930, while the North began its long decline in the early 1920s.

Occupational health was of some concern to the textile unions of the 1920s. The proceedings of the Annual Convention of the American Federation of Textile Operatives, a New England union composed mainly of skilled workers, indicate some concern with occupational health and safety throughout the 1920s. There is discussion of the inadequacies of workers' compensation legislation and of the attempt to establish the 48-hour week, but no mention of lung ailments. In 1929 the convention passed a resolution commenting on unhealthful and unsanitary conditions in textile mills and calling for their improvement, the passage of child labor laws, and the establishment of the 48-hour work week.

In the South, textile labor organizations, weakened by schism and internecine warfare, were unable to counteract employer dominance.

> Mill owners, particularly in the South, fostered a paternal relationship between themselves and their workers, many of whom lived in company-owned mill villages. Mill owners had absolute power over such workers and used that power to thwart any union activity. Suspected union members were evicted from their homes, set upon by mill-paid deputies, prevented from using the village meeting hall, hounded out of the village church, or jailed for trespassing.[11]

Nevertheless, the resurgence of the labor movement in the 1930s included the textile worker unions. By 1935, the United Textile Workers had increased its membership to almost 80,000. By 1939, the total membership of the new CIO affiliate, Textile Workers Union of America, and the remnants of the old UTW that had not gone over to the CIO, amounted to almost 85,000. Major growth in both unions was to occur with the onset of the Second World War, its massive stimulation of economic activity, and the political accommodation agreed upon by labor and management during the war effort. UTW, the AFL affiliate, was able to rebuild its membership rolls to 43,100. At the end of the war, TWUA stood at almost 250,000 (see Table I).

By 1941, in England, byssinosis had been recognized as an occupational disease and made compensable. In the United States the Textile Workers Union of America passed a convention resolution that called for

> thorough study [to be] made of the hazards in the textile industry by the United States Health Bureau with a view of enactment of Federal legislation which will eradicate health hazards for both men and women in the industry, in so far as possible.[12]

HEALTH AND SAFETY 1939–1949

Labor newspapers, published by the national organization's headquarters and sometimes by more affluent local unions, are the principal organs for communicating the economic and political agenda of trade unions to their memberships. Reports of negotiated agreements indicate the issues of greatest interest to the membership and the items *thought* to be of greatest concern by the national leadership. Key union convention decisions are explained, and the legislative agenda of the organization is presented. More than any other documents, union newspapers reveal the priorities of the organization.

Prior to World War II, *Textile Labor,* TWUA's newspaper, reported little about health and safety issues. Occasional articles reflected specific concerns—e.g., in 1939 research was reported about lung disease in the silk industry and attempts to control "rayon poison" (carbon disulphide) in Pennsylvania. In 1941 a health program at Viscose Company plants in Pennsylvania was introduced with TWUA support as a result of surveys by the Drinker brothers, a team of Harvard industrial hygienists.[13] Other examples of health improvements mentioned in the pages of *Textile Labor* in prewar times were instances of X-ray services made available for tuberculosis victims and the installation of women's facilities, drinking fountains, and cafeterias. The reports of improvements, however, were few prior to 1946. TWUA expressed concern about deplorable health conditions and occupational hazards, particularly in the South, highlighting even the slightest achievements and improvements. Health and safety, however, was not a major organizing issue for the union.

During the war years, Department of Labor statistics and U.S. Public Health Reports, excerpted and printed by TWUA, demonstrated alarming figures of absenteeism in the textile industry. According to the union, textile workers were particularly prey to ills caused by fatigue, eyestrain, lack of ventilation, and inadequate heating and sanitation. In addition, statistics were published shortly after the war on the numbers of workers killed and disabled while on their jobs.[14] From 1941 to 1944, the workplace toll was 53,000 killed and 6 1/2 million temporarily disabled. These figures presented alarming news of on-the-job sickness and accidents. Not until after the war, however, were concerns about the cotton dust hazard reflected in *Textile Labor*.

THE DRINKER REPORT

In 1944, Verne L. Zimmer, Director of the Division of Labor Standards of the U.S. Department of Labor, asked Philip Drinker, Professor of Industrial Hygiene at Harvard School of Public Health, to do a "survey of atmospheric conditions in cotton textile mills."[15] Zimmer's interest had been stimulated by requests for such a study from the Labor Commissioners of both North and South Carolina. Workers in these states had made complaints to state government about working conditions in the mills.[16] The result of the complaints (in addition to some activity at the state level) was the request to Drinker, his subsequent investigation, and publication of a special bulletin by the Division of Labor Standards in June 1945.

Drinker visited "numerous" mills in North and South Carolina with the two labor commissioners. In addition, he received "a little preliminary education" by visiting a plant owned by Nashua Manufacturing Company in

New England and consulted with textile equipment manufacturers in the North and researchers at Clemson College and North Carolina State. The document he produced provided an overview of cotton industry hazards, with the notable exception of byssinosis.

In March 1945, when Drinker was writing, the cotton textile industry was on a 24-hour schedule and the mills were "being pushed hard to increase output." Approximately 600,000 people were employed in the industry. "Rigid price control and scarcity of equipment" caused problems, but despite inadequate maintenance, "the machinery . . . continued to function."

Drinker wrote of the high level of technology in the U.S. textile industry, noting the tribute paid by a visiting British delegation. The economic pressure of the war, however, had resulted in inadequate machine maintenance and, because of increasing speed of operation, "sanitation" had been outstripped by production. "Mills which formerly were reasonably clean are today undeniably dusty." He noted the high accident rate of the textile industry, but since his study focused on atmospheric conditions, he did not examine such problems.

Drinker's report concentrated on the problems of heat, humidity, and dust control. He pointed out that "the manufacture of good yarn and good cloth requires humidity and temperature control, especially the former."

> The effects of high temperatures and humidities upon the worker have been explored very fully. There are plenty of data now available to show any skeptic that the efficiency of the worker, man or woman, begins to fall off when the dry and wet bulbs exceeds certain combinations.[17]

Drinker argued that U.S. cotton mills frequently operated "under conditions of heat and humidity which far exceed the upper limits of the human physical efficiency zone."[18] Nowhere in the pamphlet, however, did he discuss the possible health consequences of such excessive heat and humidity. He noted that air treatment was becoming more common in the industry and made recommendations for control.[19]

He did discuss the health effects of cotton dust, commenting that the industry was "not an unhealthful one," comparing favorably with the steel industry. Mill fever, he believed, was not a serious problem and "is wholly preventable by means of dust control."[20] He cited Prausnitz's British study, "Investigations on Respiratory Dust Disease in Operatives in the Cotton Industry," published by the Medical Research Council in 1936, indicating that he had heard of byssinosis.

> In this country, it seems that those who are unusually sensitive leave the industry and seek other jobs, but such job shifting does not seem to be practiced in Great Britain where a rather high incidence of asthma among cotton mill operatives is said to occur.[21]

In his conclusion, he reiterated his basic premise: ". . . health records in the cotton textile industry are good." His final word was that

> There is nothing inherently unhealthful about the industry save for its tendency to allow somewhat dusty conditions in certain processes, and hot atmospheric conditions in others.[22]

Phillip Drinker's report on atmospheric conditions in textile mills was reviewed in the January 1946 issue of *Textile Labor*. TWUA wrote, "It is no news to workers that their working conditions are uncomfortable and unhealthful." The article named cotton mill fever, also known as Monday fever, as a hazard and reported that some workers were so sensitive they had to leave their jobs. TWUA stated that although some industries had controlled their dust hazard, cotton mills had made few attempts to do so. The Union attacked the "callous indifference" of industry and suggested that industry's response to this and other health standards was long overdue.[23]

THE HARVARD COTTON PROJECT

In December of 1946, *Textile Labor* reported "Harvard Battles Diseases Common to Textile Workers":

> In experiments financed by the Nashua Manufacturing Co. and the Saco-Lowell Shops, Harvard's School of Public Health is grappling with the problem of dust and lint control in cotton textile mills to protect the workers from dangerous occupational diseases. Cotton textile workers are susceptible to mill fever and to a lung condition called byssinosis, brought about by long exposure in the opening, picking and carding rooms. Byssinosis victims are unable to perform work requiring physical exertion. Now that physicians have initially established the existence of the disease, the Harvard group, under Dr. Leslie Silverman's leadership, is experimenting with new exhaust ventilation systems to prevent it. Combined with an efficient air cleaning device, the ventilation system is designed to clear the air of dust and lint.

Byssinosis, TWUA claimed, had thus been recognized as a dangerous disease, and the union so informed its membership. The article was hopeful. Science offered the promise of prevention. Indeed, just a few months later, the Research Department of TWUA reported that "the dust house is becoming an institution of the past . . . and . . . opening and picking rooms are becoming better places to work."[24] Although, stated this report, "the last word has not been said on reducing cotton dust," improvements in exhaust systems designed by studies at Harvard School of Public Health "hope to remove all dust hazards . . . and change radically the entire preliminary processing." TWUA's Research Department presented a promising forecast to those afflicted by or concerned about the future of dust hazards.

One year later, in a similar report printed in *Textile Labor*, TWUA's Research Department outlined specific machinery to be used in the removal of dust and lint, from vacuum stripping to exhaust hoods and general room air conditioning systems. The article noted the problems faced by users of these systems, and asked union membership to notify their local union if their mills installed such systems.[25] According to *Textile Labor* then, technology was available to prevent deleterious working conditions, but the union was unsure of the extent to which this technology was actually being used.

Late in 1948, after urging textile mills to install safeguards to protect the health of workers and in so doing insure more efficient production, TWUA set out to influence state health boards to standardize and enforce correct "atmospheric conditions" in cotton mills. TWUA had produced a pamphlet entitled "Air Conditioning in Textile Mills" in which it informed workers and the general public how to determine a need for air conditioning systems and cited surveys regarding the optimal temperature for efficient production.[26] The report in the labor newspaper, however, focused more on worker productivity as affected by dusty, hot conditions than on worker health.

The manual on air conditioning in cotton textile mills created quite a stir. *Textile Labor* reported that industrial, technical, and labor circles had been requesting copies. One response to the manual by a president of an engineering firm stated that the "manual accurately sets forth temperature and humidity controls" and that "air conditioning was inevitable" and would bring increased profits. The letter argued in addition that new mills should be built without windows and that old mills should brick up their windows, but that these standards should not be legally compulsory.[27]

One of the copies of "Air Conditioning in Textile Mills" found its way into the hands of Saco-Lowell, the textile equipment manufacturer that was funding Silverman's research at Harvard.[28] The company asked Silverman to comment on the pamphlet and he did so promptly.

Silverman began his critique of the textile union report by indicating that he has known about the report "for some time." Solomon Barkin, research director for TWUA, had corresponded with Silverman eight months previously, asking for Silverman's comments, but Silverman pointed out that he "had nothing to do with the preparation of the report." The Harvard professor "tried to point out to Mr. Barkin that many of the things he was proposing were ideal and that they had not necessarily been proved in practice." Silverman was also critical of the Union's "assumption that textile workers or shop foremen—that is, Union members—are competent enough to make their own measurements of environmental conditions." Certainly in Silverman's experience, workers could not be expected "to be sufficiently familiar with technical details to make temperature and humidity surveys of this kind."

Perhaps even more interesting is Silverman on dust:

In my correspondence with Mr. Barkin I also tried to point out that they should say something about reducing dust conditions in mills, since I felt this was something which could be done quite easily and with available equipment, but they took no cognizance of my comments.

Silverman was critical of TWUA-proposed state regulation and noted that the lower rates of absenteeism in air-conditioned mills claimed by TWUA could probably not be substantiated. "Aside from its ideal aspects I believe the report may be of some value. It is questionable whether the Union people should be the ones to propose what type equipment should be installed."[29]

Saco-Lowell liked Silverman's letter: ". . . it suggests that this is a sound and thorough man who does not accept opinions of other people unless they are well supported by good factual evidence."[30]

TWUA introduced bills to standardize air quality in textile mills in Georgia in January of 1949 and in the South Carolina legislature in 1948. Georgia's General Assembly killed the bill after a few months of debate, citing the Cotton Manufacturers Association as the opposition. In North Carolina, the *Southern Textile News* of Charlotte "blasted" the state legislature for considering a bill to air condition textile mills and "accused lawmakers of straying from the path of reason . . . air conditioning would hamstring . . . strangle . . . cripple the industry."[31]

Textile labor did not have the strength in the South to win its demands politically—nor was it successful in developing labor organization and economic strength. Trade unions in cotton textiles could not hope to have direct impact on industrial technology. The long slide in membership had begun in the late 1940s.

TEXTILE LABOR AFTER TAFT-HARTLEY

The textile unions' Southern organizing success, begun during World War II, was shortlived. The watershed passage of the Taft-Hartley Act enhanced Southern textile employers' tactical arsenal for combating union organizing drives. The proportion of union representation election victories was two thirds in the period immediately preceding 1947, but fell to one third directly after the anti-labor legislation was enacted. (See Table I.)[32]

The early postwar period saw a dramatic change in textile industrial structure, the importance of which was hotly debated. Though identified by economists as one of the prime examples of a highly competitive market structure, the degree of industry concentration grew with the rise of large firms integrating mass production with mass distribution. Jesse Markham argued that the increase in scale increased industry efficiency without appreciable decrease in industry competition.[33] Solomon Barkin, TWUA's Research Director, emphasized that the increase in firm resources, particularly the increased number of both plants per firm and multiplant firms, increased company bargaining leverage over organizing workers.[34]

The rise of large, integrated firms had additional consequences for worker efforts at controlling job conditions. Another aspect of the changing firm structure was the growth of research and development departments internal to textile manufacturing firms. Given the backlog of technical innovations that were not introduced during wartime, the increase of post-World War II innovative activity led to a rapid acceleration of technical changes in production into the early 1950s. The weakened position of the unions left them less able to bargain over the productivity gains either in the form of wages or reductions in work effort.[35]

In June 1952, the *South Carolina Labor News* published in the Southern heartland of the textile industry, editorialized:

> Recent studies made in Washington, D.C., show that between 400 and 600 million man-days per year are lost because of sickness of working people. That loss is 35 times greater than the loss due to all accidents. This is another reason for the tremendous growth of the health and welfare movement in Organized Labor . . . Don't forget this: in practically all cases when a working man is sick, his pay stops![36]

This article is suggestive of a shift in labor's orientation: acceptance of the control of industrial health hazards by employers—and the desire to cushion the impact of the dangers of work via a "health and welfare" movement. The continual decline of textile labor organization drove prevention of occupational disease from the consciousness of union strategists. Between 1945 and 1955, TWUA lost approximately 50,000 members, about 20 percent of its strength, while UTW managed to hold its own. In the next decade, both unions were even more seriously damaged: TWUA fell from 197,500 to 123,000 members, and UTW from 48,100 to 36,000 (see Table 13-1). Textile labor organization was struggling for survival. Nevertheless, in 1960, for instance, the United Textile Workers *Textile Challenge* announced the opening of the Schiffli Embroidery Workers Health Center in West New York, New Jersey with great fanfare. The Center, jointly established by UTW local 211 and the Schiffli Lace and Embroidery Manufacturer's Association, would care for worker's health but included no preventative occupational health program.[37] In 1962 the TWUA convention passed a resolution calling for the improvement of worker's compensation, including coverage for "illnesses."[38] And in 1963, the UTW endorsed an AFL-CIO Labor Day Safety Campaign, urging "that each local recommend participation to its safety committee, or, if a local does not have a safety committee, that one be appointed."[39] No mention was made in either union of the disease peculiar to their memberships—byssinosis.

In 1964, however, Dominion Textiles, Ltd., a *Canadian* textile firm, in a contract with the United Textile Workers, agreed to establish jointly a "special commission, headed by medical experts" to investigate byssinosis.[40] Byssinosis was described as "a respiratory ailment long prevalent in the textile industry, which, it is suspected is caused by the dust from carding machines." The report of the findings of this study never appeared in the newspaper. The next reference to lung disease in the *Textile Challenger*, in 1967, warned of the dangers of smoking cigarettes.[41]

In 1968 the signs of an AFL-CIO campaign for an occupational health and safety law began to appear. The UTW urged support for the safety law, but did not mention byssinosis or any other particular problems of textile workers.[42] Esther Peterson, Deputy Secretary of Labor, addressed the UTW Convention and asked for its support for the pending legislation, but again no mention of cotton mill workers' respiratory disease was made. The *Textile Challenger* reported on a Government study of disabled workers, using a canned labor news service release.[43] Similarly, the TWUA Convention in 1968 was asked to consider a resolution supporting the Occupational Safety and Health bill then in Congress and urging all affiliates to negotiate the establishment of industrial hygiene programs and joint safety committees, but the measure was referred without discussion to the Executive Council for passage.

In August 1968, the first mention since the 1940s of byssinosis in a TWUA publication appeared. A delegation of TWUA officials returned from an International Labor Organization meeting and reported that "the ILO Textile Committee also urged that a study be made to combat byssinosis, an occupational disease also known as mill fever [sic], a chronic ailment which hits workers who breathe air filled with cotton dust for an extended period."[44]

In 1969, the byssinosis issue began to come to the fore. The May issue of *Textile Labor* commented at the end of an editorial on "noise in the weave room":

> But let's not lose sight of the fact that a deadly danger lurks in a cotton mill. It's the dust that pollutes the air in many carding and spinning rooms. . . . It's time the industry moved to not only muffle the noise in the weave room, but to clear the air in the cotton mill.[45]

In June the TWUA Executive Council "zeroed in on byssinosis, a lung ailment caused by cotton dust," calling for the U.S. Public Health Service "to undertake a study of the incidence of byssinosis."[46] In August *Textile Labor* published an article by Charles C. Johnson, Jr., administrator of the Consumer Protection and Environmental Health Service of the U.S. Public Health Service, entitled "Of Corpses and Corporations."

> For years, we had the comfortable illusion that byssinosis, the lung disease caused by inhaling cotton dust, was not a problem for American textile workers. Even though British workers using American Cotton came down with this

crippling lung disease, we relied on a limited X-ray survey done years ago which did not reveal a byssinosis problem.

Now, our scientists are discovering that America's 230,000 cotton textile workers are also threatened by this respiratory disease. In one mill, employing 500 people, 12% were found to have byssinosis. Thirty percent of the workers in the carding room had the disease. In another mill, 26% of the workers in the carding and spinning room were victims of byssinosis. The disease begins with "Monday morning chest tightness" and progresses to chronic bronchitis and emphysema.[47]

The following month, *Textile Labor* reported on Ralph Nader's letter to Robert Finch, Secretary of Health, Education and Welfare, urging Federal action to deal with byssinosis. Nader reported that scientists in Georgia and North Carolina were unable to gain the cooperation of industry for studies of cotton mill workers' lung disease. For the first time, the term "brown lung" was used, playing off the fairly well known "black lung" campaign of the miners. The public controversy, in which TWUA was to play an important role, had finally begun.

CONCLUSIONS

At the height of textile labor's strength, just following World War II, the dust problem in cotton mills was a matter of serious concern to the predominant trade union in the industry. The retreat (if not collapse) of textile unionism in the face of strong industry opposition and a hostile political climate in the nation as a whole, but particularly in the South, eventually swept occupational health issues from the union's agenda. Organizational survival became the top priority. Not until the broad changes in social climate of the 1960s did byssinosis reappear as an issue for organized labor. Thus, the crude power politics of labor-management struggle establishes the outlines of our answer to the questions about labor's occupational health concerns.

Economic dominance by industry, however, is not the whole story. Even at the pinnacle of union strength, lung disease was not an issue behind which the union threw its weight. "Health" was the province of the good doctors at Harvard; and trade union officialdom was convinced that the solutions to byssinosis lay in Dr. Silverman's laboratories. Until the 1970s, no American textile union had on its staff physicians or industrial hygienists who could interpret science from labor's standpoint. When Dr. Irving Selikoff addressed the TWUA Convention in 1970, the president of the union commented that this was the first time a physician had spoken to TWUA's governing body. The TWUA Research Director, George Perkel, who eventually made brown lung his own crusade, was an economist.

TWUA was one of the more progressive unions with regard to technology, in that it did have an industrial engineer in its national headquarters. And the union was willing to take up "atmospheric conditions" in the mill as a *legislative* issue in the early 1950s. The union's failure, however, to take up byssinosis as part of the air conditioning fight, even ignoring Silverman's advice, suggests the inordinate awe in which health science is held by workers and their organizations. Silverman, of course, was *not* a physician, yet TWUA characterized his research as a Harvard battle against disease. What is so pernicious about labor's dependence on the good will of the scientist is that close economic and class links tie the university to industry. The fortunate coincidence of interest between a textile *equipment* manufacturer and worker's health generated some research, but without a strong labor movement, informed and capable of taking up the issue, action on byssinosis waited another twenty years. Of course, as the political and economic fights of the early 1950s were lost by labor, ideological hegemony seems a small issue. But it was not the resurgence of textile unionism that finally placed byssinosis on the public agenda. Rather, the social movements that challenged business domination of science and technology raised occupational and environmental health as issues of concern.

Some of those same movements also challenged traditional trade union leadership. The lesson of the *black* lung movement was not only that workers could mobilize around an occupational health issue; it was also that union reform movements could successfully challenge long-entrenched leadership.[48] Even if brown lung was not an organizing issue, as some union officials believed, serious attention would have to be given to it even if only to avoid internal union political problems.

Was brown lung an organizing issue? Was trade union leadership wrong in its assessment of occupational health as an area around which to build the union? The argument rests on a number of debatable points:

1. Textile workers were poor, ill-educated, and unhealthy and were concerned about health hazards only if the effects were clear, imminent, and, perhaps, catastrophic.
2. Textile union organizers were not a lot different from the membership, and would have had difficulty in dealing with highly technical issues.
3. The clearest victims of brown lung were disabled or retired, hence not in the mill and not eligible for trade union membership.

The critique of this argument requires that we construe the labor movement in much broader terms than those of modern American business unionism. In the 1970s, the Brown Lung Association was able to build links between progressive scientists and disabled workers and to demonstrate that silk textile workers could grasp difficult issues and be a potent political force. But the Brown Lung Association never was able to organize active workers effectively. If the textile unions had embarked on a broad community organizing strategy in the 1950s, including retired workers and focusing on occupational health issues, could their decline have been averted? Would byssinosis have been dealt with earlier by industry and medicine? Such a strategy would have required earnest support and resources from the AFL-CIO, and a Southern organizing drive of substantial scope—including the development of coalitions of science and labor. Labor, on the defensive at that particular historical juncture, was incapable of such a strategy.

NOTES

1. David H. Wegman, Charles Levenstein, and Ian A. Greaves, "Byssinosis: A Role for Public Health in the Face of Scientific Uncertainty," *American Journal of Public Health* 73, no. 2(February 1983).

2. Arend Bouhuys, *The Physiology of Breathing* (New York: Grune and Stratton, 1977).

3. Richard Schilling, "Case Study No. 2: Field Studies of Byssinosis," *Journal of Occupational Medicine*, Special Supplement, October 1962; Bouhuys, *Breathing*.

4. Pennsylvania Senate Journal, 1837–38, Part 2, p. 289, cited in Anthony F. C. Wallace, *Rockdale* (New York: Norton & Co., 1978), pp. 181–82.

5. James Weinstein, *The Corporate Ideal and the Liberal State, 1900–1918*, (Boston: Beacon Press, 1968).

6. Mrs. Clara Lewis, December 20, 1981, Bath, South Carolina; quoted in Robert S. McCarl, *While I Breathe, I Hope* (Brown Lung Association, nd, np), p. 24.

7. Joseph Y. Garrison, "Textile Workers Union of America," in *Labor Unions*, ed. Gary M. Fink (Westport, CT: Greenwood Press, 1977), p. 383.

8. William Mass, "Technological Change and Industrial Relations in the Cotton Textile Industry: The Diffusion of Automatic Weaving in the United States and Britain," unpublished Ph.D. Thesis, Boston College, 1984.

9. See also William Mass and Charles Levenstein, "The Scientific, Political and Economic Factors in Occupational Health Regulation: A Case Study from the Massachusetts Cotton Industry," *American Journal of Public Health*, 1985, for a more extended discussion.

10. See Irwin Feller, "The Draper Looms in New England Textiles, 1894–1914: A Study of Innovation," *Journal of Economic History* XXV(1966):320–67; and Mass, "Technological Change."

11. Garrison, pp. 383–84.

12. TWUA *Proceedings*, 1941.

13. *Textile Labor*, November 1939 and August 1941.

14. *Textile Labor*, August 1941 and May 1945.

15. P. Drinker, "Atmospheric Conditions in Cotton Textile Plants," Special Bulletin no. 18 of the USDOL, Division of Labor Standards, June 1945.

16. Charles Levenstein, "Byssinosis: The Recognition of an Occupational Disease; A Report to NIOSH," November 1984, chapter IV.

17. Drinker, p. 6.

18. Ibid., p. 14.

19. Ibid.

20. Ibid., p. 8.

21. Ibid., p. 9.

22. Ibid., p. 17.

23. *Textile Labor*, January 1946, p. 7.

24. *Textile Labor*, "On the Job," July 19, 1947.

25. *Textile Labor*, "On the Job," May 22, 1948.

26. *Textile Labor*, "Mill Air Conditioning Made Easy," Oct. 2, 1948.

27. *Textile Labor*, Nov. 20, 1948.

28. Edward to Gwaltney, Oct. 13, 1948 (Papers of Saco-Lowell Company in Baker Library Archives at Harvard Business School, hereinafter designated SLP).

29. SLP, Silverman to Gwaltney, Oct. 22, 1948.

30. SLP, Edwards to Gwaltney, Nov. 4, 1948.

31. *Textile Labor*, "Heat Got 'Em," April 2, 1949.

32. Fink, *Labor Unions*.

33. Jesse W. Markham, "Vertical Integration in the Textile Industry," *Harvard Business Review* 28, no. 1(1950).

34. Solomon Barkin, "The Regional Significance of the Integration Movement in the Southern Textile Industry," *Southern Economic Journal* 15, no. 4 (April 1949).

35. Solomon Barkin, "Human and Social Impact of Technical Changes," *Proceedings* of Third Annual Meeting of Industrial Relations Research Association, 1950, pp. 112–27.

36. *South Carolina Labor News*, June 1952.

37. *Textile Challenger*, June–July 1960, p. 4.

38. *Proceedings*, Textile Workers Union of America, 1962.

39. *Textile Challenger*, July 1963, p. 1.

40. *Textile Challenger*, August 1964.

41. *Textile Challenger*, "Study Shows Smoking Affects Aging," October 1967.

42. *Textile Challenger*, "On-the-Job Slaughter Target of Congress," April 1968, pp. 4 and 8.

43. *Textile Challenger*, "One Sixth of Workers Disabled, Study Shows," June 1968, p. 10.

44. *Textile Labor*, "From ILO in Geneva: A Pledge to Protect Textile Workers' Rights," August 1968.

45. *Textile Labor*, May 1969, p. 10.

46. *Textile Labor*, June 1969, pp. 24 and 22.

47. *Textile Labor*, August 1969, p. 4.

48. Barbara Ellen Smith, "Black Lung: The Social Production of Disease," *International Journal of Health Services* II, no. 3(1981).

Selected Bibliography

Asher, Robert. "Workmen's Compensation in the United States, 1880–1935." Ph.D. Dissertation, University of Minnesota, 1971.

Berman, Daniel. *Death on the Job, Occupational Health and Safety Struggles in the United States.* New York: Monthly Review Press, 1979.

Brodeur, Paul. *Expendable Americans.* New York: Viking Press, 1973.

———. *Outrageous Misconduct.* New York: Pantheon Press, 1985.

Bronstein, Janet M. "Brown Lung in North Carolina: The Social Organization of an Occupational Disease." Ph.D. Dissertation, University of Kentucky, 1984.

Chavkin, Wendy. *Double Exposure.* New York: Monthly Review Press, 1984.

Gersuny, Carl. *Work Hazards and Industrial Conflict.* Hanover, NH: University Press of New England, 1981.

Graebner, William. *Coal Mining Safety in the Progressive Period. The Political Economy of Reform.* Lexington: University Press of Kentucky, 1976.

Hamilton, Alice. *Exploring the Dangerous Trades.* Boston: Little Brown and Company, 1943.

Hardy, Harriet L. *Challenging Man-Made Disease: The Memoirs of Harriet L. Hardy, M.D.,* New York: Praeger, 1983.

Nonet, Philippe. *Administrative Justice: Advocacy and Change in Government Agencies.* New York: Russell Sage, 1969.

Prouty, Andrew Mason. "More Deadly than War!—Pacific Coast Logging, 1827–1981." Ph.D. Dissertation, University of Washington, 1982.

Rosen, George. *A History of Coal Miner's Diseases.* New York: Schuman's, 1943.

Sicherman, Barbara. *Alice Hamilton: A Life in Letters.* Cambridge: Harvard University Press, 1984.

Skidmore, Hubert. *Hawk's Nest.* New York: Doubleday, Doran and Company, 1941.

Smith, Barbara Ellen. "Digging Our Own Graves: Coal Miners and the Struggle over Black Lung Disease." Ph.D. Dissertation, Brandeis University (Soc), 1981.

Szasz, Andrew. "The Dynamics of Social Regulation: A Study of the Formation and Evolution of the Occupational Safety and Health Administration," Ph.D. Dissertation, University of Wisconsin-Madison, 1982.

Weindling, Paul. *The Social History of Occupational Safety and Health.* London: Croom Helm, 1985.

Whorton, James. *Before Silent Spring.* Princeton: Princeton University Press, 1981.

Wyman, Mark. *Hard-Rock Epic: Western Miners and the Industrial Revolution, 1860–1910.* Berkeley: University of California Press, 1979.

ARTICLES

Corn, Jacqueline. "Byssinosis—An Historical Perspective." *American Journal of Industrial Medicine* 2:331–52.

———. "Historical Perspective to a Current Controversy on the Clinical Spectrum of Plumbism." *Milbank Memorial Fund Quarterly* 53 (Winter 1975), 93–114.

Derickson, Alan. "Down Solid: The Origins and Development of the Black Lung Insurgency." *Journal of Public Health Policy* 4:1 (March): 25–44.

Emmons, David M. "Immigrant Workers and Industrial Hazards: The Irish Miners of Butte, 1880–1919." *Journal of Ethnic History* 5(Fall 1985):41–64.

Fox, Daniel and Judith Stone. "Black Lung: Miner's Militancy and Medical Uncertainty, 1968–1972." *Bulletin of the History of Medicine* 54(Spring 1980):54–64.

Markowitz, Gerald and David Rosner. "More Than Economics, Federal Safety and Health Policy, 1933–1947." *Milbank Memorial Fund Quarterly, Health and Society* 64(Summer 1986).

Pratt, Joseph. "Letting the Grandchildren Do It: Environmental Planning During the Ascent of Oil as a Major Energy Source." *The Public Historian* 2(Summer 1980):28–61.

Rosner, David and Gerald Markowitz. "The Early Movement for Safety and Health." In *Sickness and Health in America*, ed. Judith Leavitt and Ronald Numbers. Madison: Wisconsin University Press, 1985.

Smith, Barbara Ellen. "History and Politics of the Black Lung Movement." *Radical America* 17(March–June 1983):89–109.

Index

Abbott, Grace: state departments of labor, 87

Accident relief plans: company-sponsored, 19–31; union-sponsored, 21–27; limitation of employer liability, 22–23; coercion of workers to join, 23–27, 28; worker resistance, 24–25; elimination of workers' right to sue for damages, 25–26, 28; economic benefits to industry, 28; public relations, 28–29; as response to workers' compensation legislation, 30; compared to workers' compensation, 30–31; as company self-insurance program, 46

Air pollution: report by Manufacturing Chemists Association, 144; study by Robert Kehoe on leaded gasoline, 144; industry-sponsorship of research, 145

Air Pollution Control Association: pro-business stance, 145

Amalgamated Clothing Workers: organizational success, 53

American Association of Industrial Physicians and Surgeons: industry ties, 151

American Communist Party: labor organizing efforts, 53

American Cyanamide Corp.: toxic chemicals and reproductive health, 166

American Federation of Labor: compensation crisis, 47; membership decline, 53; attitude toward public agencies, 55; relationship with Workers' Health Bureau, 60–61; opposition to use of lead in gasoline, 132; token involvement in radium poisoning controversy, 183

American Federation of Textile Operatives: concern with occupational health, 211

American Fund for Public Service: funding of worker health survey, 58

American Industrial Hygiene Association: industry ties, 151–52

American Medical Association: influence of lead industry, 152–53

American Public Health Association: friction among government agencies, 90; report on public health hazards of lead, 144; influence of lead industry, 152

Arrow Mutual Insurance Co.: suppression of publication of beryllium study results, 115

Asbestos: as occupational hazard, 192–205; uses and technology, 193; production and consumption, 193–95, 201; link with lung cancer, 200

Asbestos industry: sponsorship of research, 196, 197; distortion of information to workers and public, 198–99, 200; disregard for worker health, 201, 202, 205; suppression of publication of study results, 201–202; conduct of company physicians, 202, 203–205

Asbestosis: fatalities among workers, 195–96; recognized as compensable disease in England, 197; prevalence among asbestos workers, 198–99; Johns-Manville Co. policy of concealment of diagnoses from workers, 202, 203–205

Aub, Dr. Joseph: opinion on beryllium disease, 110; industry-sponsored study of lead, 148

Automobile industry: concentration of capital and effects on worker safety, 58–59; safety as issue in strikes, 63; public and occupational health issues in leaded gasoline controversy, 119–36

Avondale Mine Disaster, 71–72, 77, 82n

Baltimore and Ohio Railroad: relief association, 22

Barkin, Solomon: proposed regulation of textile mill ventilation, 216–17; effects of increasing degree of concentration in textile industry, 217–18

Beeks, Gertrude: investigation of railroad industry coercion of workers, 27; relief plan liability waivers, 28

Belknap, Elston L.: industry-sponsored studies of lead, 148–49, 153

Beyer, Clara: joint activities of Dept. of Labor and Public Health Service, 90; change in role of Department of Labor under Frances Perkins, 100n

Blum, Theodore: early diagnosis of radium poisoning case, 179

Bowditch, Manfred: association with General Electric Co., 106, 115; study of

226

Contributors

ROBERT ASHER is Associate Professor of History at the University of Connecticut-Storrs. He has written numerous articles on the history of Workmen's Compensation.

ANTHONY BALE recently received his Ph.D. from Brandeis University. He is on the editorial board of the Health PAC Bulletin and has published extensively on environmental and occupational health policy.

JACQUELINE CORN is Assistant Professor at the Johns Hopkins University School of Hygiene and Public Health. She has published on the history of occupational safety and health.

ALAN DERICKSON is a graduate student in the Department of the History of Health Sciences, University of California, San Francisco. He is currently completing a dissertation on the health programs of the Western Federation of Miners.

WILLIAM GRAEBNER is Professor of History at the State University of New York at Fredonia. He is author of a number of books and articles on retirement, social engineering and occupational health.

RUTH HEIFETZ is a M. D. and Visiting Senior Lecturer at the University of California School of Medicine at San Diego. She has been active in occupational health and worked as a physician with the Oil, Chemical and Atomic Workers International Union.

DAVID KOTELCHUCK is Director of the Environmental Health Program at Hunter College of the City University of New York. He has written articles on public policy and health and is a member of the editorial board of the Health PAC Bulletin.

CHARLES LEVENSTEIN is Professor of Economics at the University of Connecticut. He has also been a visiting scholar at Harvard University School of Public Health investigating the history of byssinosis. William Mass and Dianne Plantamura are his associates at Harvard University.

GERALD MARKOWITZ is Professor of History at John Jay College of Criminal Justice, City University of New York. He is author of a number of works in labor, social, and cultural history.

ANGELA NUGENT taught at the University of Maryland and is the author of a number of articles on various aspects of occupational safety and health.

DAVID ROSNER is Professor of History at Baruch College and City University Graduate Center. He is author of a number of works in the history of American health care.

CRAIG ZWERLING is a physician working with the United States Postal Service. In addition to his M.D., he has a doctorate in the history of science. He has published on nineteenth-century French science.

DAVID ROSNER is Professor of History at Baruch College and the City University of New York Graduate Center. He is the author of *A Once Charitable Enterprise: Hospitals and Health Care in Brooklyn and New York, 1855–1915* and editor with Susan Reverby of *Health Care in America: Essays in Social History*.

GERALD MARKOWITZ is Professor of History at John Jay College of Criminal Justice. He is co-author with Marlene Park of *Democratic Vistas: Post Offices and Public Art in the New Deal* and editor of *The Anti-Imperialists*. Markowitz and Rosner are also editors of *"Slaves of the Depression," Workers' Letters about Life on the Job*.